W. Kitchen Parker

On Mammalian Descent

The Hunterian Lectures for 1884

W. Kitchen Parker

On Mammalian Descent
The Hunterian Lectures for 1884

ISBN/EAN: 9783337160364

Printed in Europe, USA, Canada, Australia, Japan

Cover: Foto ©ninafisch / pixelio.de

More available books at **www.hansebooks.com**

ON MAMMALIAN DESCENT:

THE HUNTERIAN LECTURES
FOR 1884.

BEING NINE LECTURES DELIVERED IN THE THEATRE OF THE ROYAL COLLEGE OF SURGEONS DURING FEBRUARY, 1884.

BY

W. KITCHEN PARKER, F.R.S.,
HUNTERIAN PROFESSOR, ROYAL COLLEGE OF SURGEONS OF ENGLAND.

With Addenda and Illustrations.

LONDON:
CHARLES GRIFFIN & COMPANY,
EXETER STREET, STRAND.
1885.

[*All Rights Reserved*.]

TO

MISS ARABELLA BUCKLEY
(*Mrs. Fisher*)

This little Work is Inscribed

AS A SLIGHT MARK OF REGARD AND ESTEEM
BY HER FRIEND

THE AUTHOR.

PREFACE.

For many years past I have been engaged in researches into the structure and development of the vertebrate skeleton, and the results of these investigations have, from time to time, been published in the Transactions of various Scientific Societies—naturally, in technical language. The Hunterian Lectures also of former Sessions, which preceded the present course, were delivered in the terms of Biological Science, and were thus, of necessity, unintelligible to persons not familiar with studies of this kind.

During the last year or two, however, many useful suggestions with regard to a more popular method of treating these matters have been made to me by my esteemed friend, best known to the reading world as Miss Arabella Buckley. To her I am indebted for the plan of this last course.

The following Lectures, therefore, I offer to my non-scientific friends as slips and cuttings from the Biological Tree, in the belief that the great and absorbing question with which they deal—the problem of "Man's place in Nature"—will be found to invest even minute details with a very real interest.

And let me observe, in passing, that the doctrine of the gradual development of organic types, if it does not stand or fall with Embryology, yet must look for its greatest support from, or be contradicted by, that most important Science—the true root-stock of Biology. There is, however, no branch of human knowledge that is so difficult to put into language which can be appreciated by those who are not familiar with its special methods, its facts, and its descriptive terms.

One thing I cannot pass over without remark, and that is, the strong and almost insuperable *a priori* objection in many minds to the deductions of modern Biology; this must neither be lightly overlooked nor treated flippantly. To these opponents the biologist may say:—"It is a very light thing that I should be judged of you, or of man's judgment";—and yet he is pained at the thought of even *seeming* to be in opposition to much that the greatest and best minds hold sacred.

The biologist having given expression to this feeling on his part, it is certainly the duty of the non-biological opponent of his deductions to look these things fairly in the face; the burden of disproof is now laid upon the objector.

In conclusion, I desire to take this opportunity of thanking very heartily those friends who have kindly helped me to see the work through the press.

W. K. PARKER.

London, *November* 1884.

TABLE OF CONTENTS.

LECTURE I.
Introductory.

	Page		Page
The New Things of Biology,	1	Care of the Offspring,	13
Time, the Sexton,	2	The Crown of Creation,	14
Duckbill and Echidna,	3	Larviform Embryo of the Mole,	15
Groups of the Mammalia,	4	Great Weights on Small Wires,	16
Charles Darwin,	5	Influence of Surroundings,	17
The first Historical Man,	6	Adaptation or Extinction,	18
The Australian Region,	7	Metamorphosis of the Frog,	19
The first Creature with Nails,	8	Man's Ancestors all buried,	20
Primary Amphibia,	9	Eastern Cosmogony,	21
Protozoa and Metazoa,	10	Hyperbolical Asiatics,	22
Low Ancestral Types,	11	A Buddhist Miracle,	23
Family Arrangements,	12	Antiquated Theories,	24

LECTURE II.
Prototheria (Monotremata).

	Page		Page
The Duck-billed Platypus,	25	Compound Nature of the Lower Jaw,	42
Negative Evidence,	26	Bones forming the Ear-chain,	43
The Ceratodus a Gift of Providence,	27	New Elements of the Ear,	44
The Lancelet and Sea-squid,	28	The Egg-breaking Beak,	45
The Axolotl,	29	The Frog's Kindred,	46
Lower parts of the Earth,	30	Ancestors of the Mammalia,	47
Mammalian Advance,	31	Young Duckbills,	48
The Prototheria quasi-Reptilian,	32	All Animals come from Eggs,	49
Shoulder-bones of Duckbill,	33	Evolution of a Hairy Creature,	50
Three Collar-bones in Monotremes,	34	Relation of Mammals to Reptiles,	51
		Jacobson's Organs,	52
Extinct Birds,	35	Old and New Structures,	53
Feeling the Way upwards,	36	A Country without Birds,	54
The Duckbill's Skull,	37	Skeleton of the Prototheria,	55
Opening a new Scroll,	38	Primary Respiratory Organ,	56
Hingeing of the Lower Jaw,	39	Respiration of a Soft-shelled Turtle,	57
Surgery or Horticulture,	40		
Formation of the Lower Jaw,	41	Oxygen consumed in Pharyngeal Respiration,	58

LECTURE III.

ON THE MARSUPIALS, OR POUCHED ANIMALS (METATHERIA).

	Page		Page
A Supposed Common Root-stock,	59	A Bye-path Meadow,	74
Embryonic Membranes,	60	The hard palate,	75
Embryo of Opossum,	61	A Semi-Marsupial Insectivore,	76
New-born Kangaroo,	62	Crocodile, Bird, and Opossum,	77
Early Life of a Marsupial,	63	The Turkish Saddle,	78
Mighty Hunters,	64	The Stirrup and Columella,	79
Feeding-Grounds,	65	The Arch of the Tongue,	80
The Earth's Shaking Fits,	66	Mimetic Types,	81
Pouch-bearing Animals,	67	Works on the Marsupials,	82
An opposable Inner Toe,	68	Fossil Marsupials,	83
The Skull in Marsupials,	69	An Oak Tree,	84
Skull of Embryo Pig,	70	The Australian Flora and Fauna,	85
Ear-drum of Marsupials,	71	Frogs and their Progeny,	86
Exit of the Optic Nerve,	72	Newts and Frogs,	87
Links in the Chain of Life,	73	Nourishment of Early Germ,	88
		Development of the Germ,	89

LECTURE IV.

EDENTATA.

	Page		Page
Distribution of the Edentata,	90	A new Hinge for the Lower Jaw,	108
Specialisation of Low Types,	91	Nature's Amputations,	109
Nature's dull Children,	92	Structure, Habits, and Classification of the Edentata,	110
The Struggle for Life,	93		
Dwarf Edentata,	94		
The Extinct Glyptodon,	95	Papers on the Edentata,	111
Hair turned into Armour,	96	Ant-eaters and Pangolins,	112
The Aard-Vark,	97	The Captive Pangolin,	113
A stunted Genealogical Tree,	98	Strong Evidence for Darwinism,	114
Isolation of the Edentata,	99	Gigantic Sloths,	115
Variation of the number in the Neck-Joints,	100	Ant-eating Birds and Lizards,	116
		Food and Safety,	117
Dying out of the Teeth,	101	Love of Offspring,	118
When the Vertebrata had no Lungs,	102	Mother Carey,	119
		An Unclean Sacrifice,	120
The Hard Palate,	103	Southey as a Darwinian,	121
The Deep Things of Morphology,	104	The Green Turtle and the Glyptodon,	122
Skull of Sloth and Ant-eater,	105		
Toothless Jaws,	106	Plucking up Cedars,	123
The Feeble-faced Pangolin,	107	The Grave of the Giant,	124

LECTURE V.

INSECTIVORA.

	Page		Page
Materials for Work,	125	Wasted Types,	130
Edentata and Insectivora compared,	126	Adaptive Changes,	131
		Ungulate Lemurs and other Generalised Types,	132, 133
The First Beast-namer,	127		
Development of the Whale and the Bat,	128	The Number Five in Digits,	134
		Plastic Types,	135
Organic Root-stocks,	129	Layers of the Skull,	136

CONTENTS.

	Page		Page
The Hedgehog's Skull,	137	Arches of the Face,	141
Skull of Cartilaginous Fishes,	138	Primitive Types of Mammalia,	142
Ear-drum of Hedgehog,	139	Bibliography,	143
Parts of the Palate and Ear in Hedgehog's Skull,	140	Fossil Horses,	144
		Extinct Mammals,	145

LECTURE VI.

INSECTIVORA—continued.

	Page		Page
The Mole very Archaic,	146	Diet of the Mole,	156
Greediness of the Mole,	147	Red Teeth of the Shrew,	157
Blind, but Quick of Hearing,	148	Young of the Shrew,	158
Development of the Mole,	149	Skull of Shrew,	159
Records of Lost Types,	150	Young of Shrew like Eocene Mammals,	160
An Hereditary Agriculturist,	151	Distribution of the Shrews,	161
The Formation of the Tissues of the Mole,	152	Browzing and Grazing,	162
A Dry Skull,	153	Improvement of Breeds,	163
Air-Galleries of the Ear,	154	*Homo diluvii testis*,	164
Labyrinth of the Ear,	155	Signs of Transformation,	165

LECTURE VII.

INSECTIVORA—concluded.

	Page		Page
Foreign Insectivora,	166	The Colugo and the larger Bats,	178
Mascarene Types,	167	The Basi-Cranial Beam,	179
Skull of the Tenrec,	168	Relations of Air-Breathing Types,	180
Nasal Labyrinth of the Tenrec,	169	Old Remnants in New Types,	181
Lesser Kinds of the Centetidæ,	170	The Rhynchocyon,	182
Crested Skull of the Tenrec,	171	A Proboscidean Insectivore,	183
Wallace's Line,	172	Marsupial Characters,	184
A Complete Orbital Ring,	173	Working out old Strains,	185
The Young Colugo,	174	Bibliography of Insectivora,	186
A Puzzling Type,	175	New Insectivora,	187
A Tertiary Bat,	176		
Comb-like Teeth of Colugo,	177		

LECTURE VIII.

THE REMAINING ORDERS OF MAMMALIA.

	Page		Page
The Chiroptera,	188	Zoological Position of Lemurs,	196
A Long-fingered Animal,	189	Diet of the Aye-aye,	197
Evolution of a Flying Animal,	190	The Carnivora,	198
An Unfolded Bat,	191	Aquatic Mammals,	199
The Rodentia,	192	The Quadrumana,	200
Voyage of a Naturalist,	193	The Pig and the Ruminants,	201
Skull of a Guinea-Pig,	194	Evolution of the Horse Type,	202
Lemurs and Aye-aye,	195		

xii

CONTENTS.

LECTURE IX.
Conclusion.

	Page		Page
Metamorphosis of the Dragon-fly,	203	Searchers for Facts,	216
		Dry Light,	217
The Possibilities of Being,	204	A new Atlantis,	218
Soul and Body,	205	Quotations from Bacon,	219
Human Longings,	206	Second Causes,	220
Waves and Pulses,	207	Gentle Modifications,	221
Arrested and Developed Types,	208	A Tithe-Farmer,	222
Our Forefathers,	209	Goethe on the Teleologists,	223
Astronomy, Geology, Biology,	210	The Seal,	224
The Links of Created Being,	211	Time-Marks,	225
Our Growing Knowledge,	212	Plants and Animals,	226
The Currents of Protoplasm,	213	How the Bones Grow,	227
Embryological Work,	214	Exogenus Growth,	228
Written on the Brain,	215	Expansion of Modern Thought,	229

MAMMALIAN DESCENT.

LECTURE I.

INTRODUCTORY.

BRIEF as is the title I have given to the present Course of Lectures, it contains enough, in two words, to give alarm to cautious and timid minds.

I need hardly say that no harm is intended by it; and I believe that no harm will happen to the mind of any one who will listen to me whilst I bring forward some of the "new things" of Biology.

The Mammalia are important to the biologist beyond any other group whatever; for they contain, within their circle, the highest known type of living creatures. A group which culminates in Man may well deserve our attention and study; even the forms that make the nearest approach to the human race are, of necessity, full of interest to us.

If in human society the toe of the peasant now and then galls the kibe of the courtier, so in this class the toe of the irrational beast treads, in some cases, very close upon the heel of rational man.

It is worthy of remark that this keeper of the mammals—himself a mammal—is, *in size*, a good practical mean between the extremes. On one hand we have "the smallest monstrous Mouse that creeps on floor," and on the other the unwieldy Whale—large as an island.

Taking the class as a whole, as to intelligence, we have "the extremity of both ends"—Man at one end, and the frog-witted *Duckbill* at the other.

The mammalian class is, indeed, a most motley assemblage, whether we consider their form, their size, or their intelligence, for the wars of time have sadly thinned the ranks of regiment after regiment. Nor is the Darwinian in fault, if, when the roll-call is made, so few are found to respond to it.

Let those who clamour for connecting-links lay this to heart: myriads upon myriads of mammals have perished in the struggle of life; time has buried them. "Time, that old clock-setter, that bald sexton, time," has much to answer for; he has not only thrown the mould over the links of the chain of types, but—to change the image—he has buried thousands of complete family trees, and the geological miner has only unearthed a broken twig here and there.

He is in a low state of mental development who is unaware of the extreme antiquity of the planet on which we dwell; and it is a far cry backwards, to the time when the young of four-footed beasts *first tasted milk*.

The group, class, or family—as we may call it—which acquired the peculiar faculty of giving of their very substance to their offspring, is as ancient and as venerable as the group of the reptiles, out of which arose the feathered tribes.

Not out of the *stem*, however, of the reptilian family tree, but out of its *root-stock;* and close to that fine sucker, there shot up this other branch, to become the new life-tree of the hairy creatures, that give their young ones suck. Two of the first twigs of that new shoot are still represented by the "Monotremes," namely, the *Duckbill* and the *Echidna;* but, of course, as their line of ancestors must have existed during the formation of the outer half of the earth's ribs, they have had time enough for much specialisation in their structure. Therefore, the scientific imagination, after assuring itself that these living waifs do not lie at the root of mammalian being, bodies forth much lower and more generalised milch-kine than them.

There are fossil remains, evidently mammalian, from the base of the Secondary rocks. Whether these small jaw-bones belonged to Monotremes that had teeth, or to the more ancient Marsupials,[1] does not affect our argument. Mammalian remains will, I feel sure, turn up some day from older rocks: anyhow, in certain strata

---

[1] The *Monotremes* are so called because they have only one common outlet to their body, as in Reptiles and Birds; *Marsupials* are so called because they possess a *marsupium* or pouch.

of the Secondary epoch we get various remains of the group above the Monotremes—the Marsupials.

Professor Huxley's classification of the mammals is as follows :—

1. *Prototheria*, or the Monotremes : examples—*Duck-bill* and *Echidna*.

These are the lowest mammals known; they have udders, or milk-glands, but no teats, and in many things stand on the same level as the Sauropsida (reptiles and birds).

2. *Metatheria*, or the Marsupials : examples—*Opossum*, *Phalanger*, *Kangaroo*. These have, besides the milk-glands, perfect teats, but their young are born so early that they derive no direct nourishment from the mother until they are placed on the teat.

2. *Eutheria*.—These forms are the highest, and their young do derive direct nourishment from the mother for a considerable time before birth—before they are nourished by milk. In this group we have *Moles* and *Men*, and all the forms that lie between these two extremes. I shall speak of the Mole as a low Eutherian, of Man and of his Horse as high Eutheria.

There are at present three groups of labourers working at the Mammalia ; as, indeed at other types also ; these are :—

1. The *Zoologists*. These study the finished form, habits, and distribution of the various types, in the *present* state of the planet.

2. The Palæontologists. These study the fossil remains of the extinct forms, and their *past* distribution.

3. The Embryologists. These men are working out the development of this or that type, following it through the various stages of the history of its life.

These three divisions of the swarm of biological bees are rifling the treasures of this planet, "which pillage they with merry march bring home." They are all Darwinians, to a man; and they scout the "lazy, yawning drone" who eats of their honey, buzzing the while dissatisfaction at their work and their song.

This seems, in the ears of many, to be a "new" song, but it is, indeed, the old song spoken of by that fine old eastern naturalist—the much-suffering Job.

Who the morning stars of science were, we know not; the voice of one who lately spoke to us, in his wisdom, of living creatures, vegetable and animal, is, to our sorrow, now silent.

There is a growing consensus, or harmony, amongst the three main divisions of the workers, who are now beginning to understand each other. This has taken time, for the harmony was not, at first, in the mind of the workers, but in Nature herself; they were working apart—each group apart, and each labourer apart; but the zoological scouts, the earth-diggers, and the miners of the organisms, all these work well together now.

I think that he who digs down, so as to see Nature, alone, at her work, in the (figuratively) lower parts of the earth, will reap the richest reward, and have the highest honours.

For, the palæontologist, in his luckiest finds, brings you up, at most, the framework, only, of a creature, and that, mostly, in its adult condition. Such shreds and patches of old organisms as he finds, although unspeakably precious, are difficult materials out of which to construct a history. But your embryologist has learned where to find little compendiums of the past history of the folks that did live here a long time ago. These, however, are written, so to speak, in shorthand, and are as difficult of interpretation as the old cuneiform characters; that difficulty at once whets his ingenuity, and also makes him resolve to possess his soul in patience.

You will see that I am weaving this web of words to catch your attention whilst I bring in my theory—not mine, but Darwin's,—and yet *mine*, notwithstanding.

This theory seems to contradict the Sacred Records; it does contradict the letter of certain passages—taken alone. Man was created perfect—that is, "calling the end from the beginning;" and in the fulness of the times a Perfect Man did appear. We have His history. Having that, I, for one, care not a jot about any further history of the weak man who blots the historic page at its very beginning.

Now let us leave Man, and go down among the beasts; they are delightful creatures,—"what God hath cleansed" [made pure and beautiful] "that call not thou common."

Of the Prototheria (first beasts), the lowest, teatless mammals, we have, now, only two types (genera) left. These are both limited, in their range, within the Australian region; they are, the *Ornithorhynchus*, and the *Echidna*. The former is the lower of the two kinds; but Professor Huxley's conception of the group in its early, and perhaps abundant, existence is, that it was composed of much less specialised forms than those now living.

Are we to stand like men who cannot find their hands, because Nature and Time have buried nearly all the truly old families of the Mammalia? If we are unable to frame convenient hypotheses, to be used as intellectual scaffolding to our facts, we are out of our place in attempting biological research. Let us, if such be the case, stand out of the sunshine of fitter and abler men.

At present, I have only partially worked out the young of one of these kinds,—the *Ornithorhynchus*; but although tolerably familiar with the structure of the Vertebrata generally, I am at a loss, even in this early stage of research, to see the meaning of many things in that type.

Here is a beast—a primary kind of beast, a *Prototherian*—whose general structure puts it somewhere on the same level as low reptiles, and old sorts of birds; but in which there are characters much more archaic than anything seen in Serpents, Lizards, Tortoises, Crocodiles, or in Emeus. Therefore the existing reptiles and birds must stand aside as having nothing to do with the family tree of the Monotremes, although in some things they are like these beasts, and many of their organs are formed on a similar pattern; they are all equally below the morphological level of the nobler Mammalia.

Although some of the mammalian characters are well marked out in the Monotremes, yet they only agree with the higher Mammalia to a very limited extent; some things now seen in them are quite new to me.

Most of the existing fishes have ceased, or nearly ceased, to begin life in a larval form, or one lower than that in which they spend the main part of their existence. Some degree of metamorphosis is seen in the Ganoids, and more in the Lamprey; but with the Amphibia (Newts and Salamanders, Frogs and Toads), it is in the larval, or *grub-stage*, that we find the oldest things. One kind of frog, from the Cape, is the first creature that is possessed of nails to its toes and fingers; it is termed *Dactylethra*. The tadpole of this frog has shown me some of the most archaic structures, and the lowest condition of the tissues themselves, that I have met with in the whole sub-kingdom of the Vertebrata.

I speak of this for two reasons:—first, I find things in it which are quite like what are to be seen in the

Duckbill, but which I have in vain sought for in any other type. My second reason is this, namely, that the extremely wide range of structure taken by each individual of that species, and indeed, of all its kindred, from the time of hatching to the time when the permanent adult form has been reached, is such as to suggest almost limitless possibilities in the development of the Vertebrata, so that my thoughts run almost unconsciously parallel with the suggestions so ably put by Professor Huxley, in his paper. I can, and do imagine a group below the Prototheria, their root-stock, which may well be called "Hypotheria," or creatures *under* the beasts.

That these were akin, closely akin, to the *primary Amphibia*, there is every reason to believe. If they were metamorphic, and that I think is very probable, they lived in their infancy in the water, and their respiration was *aquatic*.

Our present work, however, is not to stand peering down into those dark depths, but to see whether the stages of the *existing* Prototheria will not show us many instructive facts. Yet even here we are almost as poor in embryos of these types, as we are in their fossils; and the present destruction of these invaluable types, for the sake of mere museum exhibition, painfully suggests their probable early extinction.

The Royal Society, however, has lately, with great liberality, furnished certain scientific *Knights* with means for following this quest—that of finding the embryos of the Monotremes. If they succeed, and we can get the early stages of the existing Prototheria, I have no doubt but that we shall be able to see much further into "that dark backward and abysm of time," when the Huxleyan Hypotheria did duty for the existing Mammalia.

I shall show in these lectures that some of the lower kinds of Eutheria (placental mammals) undergo, in their pre-natal state, and also during their infancy and youth, most remarkable transformations. I use the word transformation in a popular sense, as the term "metamorphosis" has a very limited and absolute meaning in science.

All the animals above the Protozoa (first creatures) are called "Metazoa," because they undergo remarkable changes of form, *beyond* their first stage or state. When these various stages are gone through in the active condition—the partially developed animal leading for some time a free and out-of-door life—it is said to undergo *metamorphosis*. If these changes are not utilised, if they are *pre-natal*, and the new-born active creature is practically the same as the adult, the more familiar word *transformation* is employed for the unused early changes.

A Snake undergoes remarkable *transformation* whilst in the egg; a Frog is marvellously *metamorphosed* during its active life.

The early transformations of those types which have no larval stage I look upon as the unused equivalents of the metamorphic steps of types, which, like insects, have an active larval stage, or stages. These transitory, unused stages are, manifestly, of an *historical* import; they suggest to the Darwinian lower and still lower types of ancestral animals—the Fauna of a bygone time. And this view of the matter is well borne out by what we already know of the structure of the Prototheria, or Monotremes, and of the Metatheria, or Marsupials. It is also borne out by everything I have seen, as yet, in the structure and development of the other groups of the Vertebrata, as they rise one above the other in the order of morphological excellence.

The perfection of every organ for its special use in the adult makes it the more noteworthy that there are so many things to be found, during growth, that are not only useless, but, as a rule, transient—some of them, however, are permanent. The doctrine of Final Purpose, on its old platform, and taken as if it were the conclusion of the matter, wholly fails to explain these rudiments, or remnants—these useless odds and ends of forgotten organs. Whether transitory, or permanent, they are the opprobrium of the teleologist who has not studied the growth of the embryo, but they are goads and spurs to him who devotes himself to the study of Development. Yet the doctrine of Final Causes is not affected by this deeper kind of research—it is merely placed on a higher platform.

Looking upon these facts as having an historical meaning, we shall find some light thrown upon many unexplained problems in the structure and development of our own frame.

If it be a fact that these unused and useless things were once useful, and did their work as part of the machinery of a living and active, but low form, and that in conformity with new conditions this *low, larval* life became abbreviated, then some light is shed upon these dark problems. Let this once be well settled upon sure foundations, or as a strong nail in the wall of science, and then we shall be able to harmonise our old faith with our new knowledge.

The study of the lower forms is ever giving us fresh and fresh evidence of the infinite fertility of nature's expedients in adapting living forms to varying conditions—to new surroundings. And this is seen more especially with regard to the Family Arrangements of the various types, in the matter of *paternity, maternity,* and *time.*

The surroundings are often cruel and destructive; yet nature does not temper the weather to the lamb—she clothes the lamb so that it can brave the weather.

All through nature we see that the most marvellous care is taken so that the Children shall have their chance; mostly the care is on the maternal side, in some cases, nature teaches even the stupid husband to do his part in wrapping up the children, and in keeping the home safe.

But the most wonderful part of the family arrangements takes place before the period of hatching, or of birth, the parents being unconscious agents. In many cases the parents are workers together with nature in preserving the germ; albeit the casket of this treasure is wholly unlike, as an egg, the infant that is to be hereafter in their own image and their own likeness.

It is not in the human kind, but among the cattle, that the young one is made to do most of its development in the dark, so that at birth it is strong and in good liking. This is the very culmination of reproductive adaptation; the furthest from the careless, thoughtless state of things seen in low, fish life, where, as in the cod-fish, millions of germs are sown broadcast upon the waters by one mother, who is hardened against her young ones, as though they were not hers.

But the growing care of the germ by the proper living mother, she hiding it longer and longer in her bosom before she commits it to the waters, is well seen in certain sharks, that do, in the most striking manner anticipate the last and most perfect specialisation of this kind.

The family, as such, both in birds and mammals, is not seen in its perfection among those creatures whose young are ripest at the time of birth. Birds are divided by some ornithologists into "Præcoces" and "Altrices;"

it is among the latter, whose young are feeble at the time of hatching, that the most tender care is taken of the family. The same holds good amongst mammals, as we all know; yet the young one, when born in a tender state, has fairly the form of its parents; it does not shock them by its larval ugliness; and the first human mother on record, seeing her first-born son, exclaims—" I have gotten a *Man* from the LORD."

All these things are so familiar to us that I fear you will wonder why I speak of them; but if you reflect upon the *way* in which they came about, and the *time* they took to get perfected, you will see what I am aiming at. Mentally, in imagination, I have been tarrying Nature's leisure, whilst, during untold ages, she has wrought all these wonders.

If anyone will consider the great uniformity, both in size and shape, of all mammalian embryos and germs, he will see that the marvel of evolution is always going on in a thousand types, here, in the highest class, at the very top and crown of creation. That which is now, is like that which has been; the mere shortness or length of time during which the various processes of growth and development take place is a non-essential matter. The embryo of a mammal at the stage which represents a gill-bearing vertebrate, in all cases that I have examined, ranges from one-third of an inch to an inch in length; the former size belongs to the smaller kinds, the latter to the larger.

Know one, know all; one diagram would represent all, one description serve for all.

Such a stage, moreover, gives us a form extremely like that of any other *gill-less* type—bird or reptile; while to make it into a semblance of the lower aquatic types, more "visceral arches," with more and more gaping clefts, are all that would have to be added.

In all we have the curved, larviform creature, with its

Fig. 1.—Embryo of Mole (*Talpa europaea*, 1st stage), magnified 12 diameters.

large brain-lobes bent under, in front; its tail-end bent under, behind; its solid front folds, its rudimentary gill-openings, and its paddle-shaped limb-buds. But the characters derived from its more immediate ancestry soon show themselves. By the time the gaps in the throat are filled up, and the embryo has doubled its length, the special characters of the type to which it belongs begin to be seen.

Nevertheless, in the embryo of a medium-sized mammal—*e.g.*, the Hyomoschus, a generalised ruminant—nearly an inch long round the curve, I find nothing that suggests its proper place in nature : it might belong to a Lion, or to a Gorilla, as far as its outer form is concerned. And yet an embryo of this kind—a sort of temporary, sleeping, dependent larva—becomes, in one case, a Rear-mouse with leathern wings, and in another a Whale, whose skin and blubber are as thick as a house-wall.

Lord Bacon gravely remarks that—"God hangs the greatest matters on the smallest wires." He might have been an embryologist; certainly, neither a Darwin nor a Huxley could have put such an aphorism into a better form.

Whale's eggs are no larger than "fern seed;" and yet the protoplasm in any one of them has the power, when planted where it can get due nourishment, to develop an embryo which, whilst as yet it is unborn, is as large as a good-sized cow. This phenomenon of development, which is always repeating itself in all mammals, only not to so huge a bulk as in the instance just given, is as great a "sign" or "wonder," or "miracle," as anything suggested by the most thorough-going Darwinian as part of the process of secular evolution.

All this differentiation, all this development of complex, correlated organs, in one single organism, worketh that one and the self-same force, bringing forth severally, according to the ancestry of each, modified and fashioned into various types during untold ages of the past, the various Mammalia that tenant the waters, flit in the air, or trample the paths of the forest. In each of these the *force* is manifestly the same, essentially; but the *surroundings* of the organism in which this force has been enshrined have been the same during no two successive moments of time, during all the ages in which the earth has brought forth living creatures.

The sensitiveness of a living creature to outward impressions is excellently put by our great poet. He says that you cannot press your hand with a rush, but it will bear a visible mark or cicatrix, and that the eyes do shut their coward gates on atomies.

The infinite number of delicate and gentle modifications in the human form, all speak eloquently of the influence of "surroundings." All the races of this type are evidently varieties of one common species; a species whose existence upon this planet, *according to Usher*, has been barely six thousand years. As the wind pipes, so the creatures dance; and the wind and the sun are ever renewing their old contest as to who can make the traveller pull off his cloak first.

For a long while the eager, nipping wind of Siberia tried this on with the *Mammoth:* he merely had his cloak made warmer and thicker. The wind ultimately killed the beast, but never got him to take his garment off. On the other hand, in the tropics of the Old World, to this day, the brothers of the old Mammoth have been living in harmony with the Sun; but they have thrown away their cloaks, and bask, naked, in his beams.

But, during great, sudden changes in the home or feeding-ground of animals, the dilemma has again and again been *adaptation* or *extinction*; in many cases nothing short of *metamorphosis* has saved them from death, and kept them alive in famine.

Speaking of metamorphosis, I am brought to that which is, evidently, the key to the intricate wards of this long-locked-up problem—I mean the descent of organic types. The metamorphosis of insects—a marvel always fresh and wonderful both to the man of years and to the child—reveals to us the practically infinite possibilities of the modifications that may take place in the lifetime of a single worm-like creature.

If we were not thoroughly familiar, from our childhood, with the astounding phenomena of insect-transformation, if we only knew the Grub, the Pupa, and the perfect, winged Imago, separately, any assertion of such a possibility by some far-seeing biologist, would be treated with contempt, and the brand of heresy would be set upon him.

Such a developmentalist would fare as Bruce, the traveller, fared, when he related his adventures, telling of the sights his eyes had seen—

"All he gets for his harangue is—'Well!
What monstrous lies some travellers tell.'"

How does all this bear on mammalian descent? Mammalia are not insects.

My answer to this curt but pertinent question is—that insects show us what is possible as to metamorphosis in a very high group of the Metazoa, or creatures that change their form during their development. Now, as I have spent the spring and summer, and some part of the grey autumn of my life in observing the phenomena of metamorphosis in the Vertebrata, you will, I hope, of your clemency, listen to my words.

Before Darwin's *Origin of Species* had for any length of time been printed and discussed, I had seen such things in the metamorphosis of the common Frog as seemed to me like the writing in a newly-opened scroll of science. Starting as one of the lowest and most generalised kinds of fish, this creature does not end his strange, eventful history until he has given us the type and promise of almost everything in the structure of a high mammal (or Eutherian), even of Man himself, who lifts himself up above his mammalian fellows.

Long before our era a gifted captive Jew saw, among the celestial hierarchies that appeared to him in vision, "the likeness as the appearance of a man." To us it is given to see man's image down among the living creatures that crowd around the foot of Jacob's ladder. Bacon remarks that—"Light doth stream down more clearly and divinely into the mind of a young than of an old man, for it is written—'Your old men shall dream dreams,' but 'your young men, shall see visions.'" Now, if one of the bright young soldiers in our rapidly increasing army were permitted to see the whole web (woof and warp) of organic life, he would everywhere see glimpses of the human face divine; the features of the *latest* creature would be traceable in the face of the *earliest*.

Yet these types and foreshadowings of the great Reasoner, to be developed in the parturient fulness of time, only reached their own little Pisgahs; they looked over towards the human territory, but they entered not in. As for the direct ancestors of man, time has buried them, and no man knoweth of their sepulchre to this day.

ADDENDUM TO LECTURE I.

That which is biological in the foregoing Lecture will be considered and treated of from time to time in the succeeding Lectures, and also in the Addenda attached to them. But there is one thing that may be brought in here, namely, the conceptions that the Ancients held with regard to the Origin of the Universe, and especially of living creatures. Amongst these the Jewish Bards stand first, far in front, indeed, and moreover their poems have been worthily rendered into what Swinburne truly calls "Divine English."

I am, of course, well aware that Moses, and Job, and David were not the only great and wise and good men who in ancient times sang —"How the Earth rose out of Chaos."

Whilst composing these Lectures, a friend kindly put into my hands two invaluable works that have yielded me great pleasure and profit. The first of these is *A Manual of Buddhism*,[1] by R. Spence Hardy;

[1] 2nd edition. London: Williams & Norgate. 1880.

the second is *Religious Thoughts and Life in India*,[1] by Professor Monier Williams, M.A., C.I.E. Part I., Vedism, Brahmanism, and Hinduism.

Some of the venerable hymns given in the latter work are sublime, and certainly come very near the Hebrew writings.

I have only space for a specimen or two, but trust that the reader will be led to avail himself of Professor Williams's invaluable labours:—

"In the beginning there was neither nought nor aught;
Then there was neither sky nor atmosphere above.
What then enshrouded all this teeming universe?
In the receptacle of what was it contained?
Was it enveloped in the gulf profound of water?
Then was there neither death nor immortality;
Then was there neither day, nor night, nor light, nor darkness.
Only the Existent One breathed calmly, self-contained.
Nought else but he there was—nought else above, beyond.
Then first came darkness hid in darkness, gloom in gloom;
Next all was water, all a chaos indiscrete,
In which the One lay void, shrouded in nothingness.
Then turning inwards, he by self-developed force
Of inner fervour and intense abstraction, grew.
First in his mind was formed Desire, the primal germ
Productive, which the wise, profoundly searching, say
Is the first subtle bond, connecting Entity
And Nullity."—Page 13.

"From glowing heat sprang all existing things,
Yea, all the order of this universe (*Rita*).
Thence also Night and heaving Ocean sprang;
And next to heaving Ocean rose the Year,
Dividing day from night. All mortal men
Who close the eyelid are his subjects; he
The great Disposer, made in due succession
Sun, moon, and sky, earth, middle air, and heaven."—Page 404.

[1] London: John Murray. 1883.

I also give a few quotations from Mr Hardy's very important work, but these are to illustrate what is to me most remarkable, namely, the manner in which these Asiatic people threw the reins on the neck of their imagination.

I cannot but think that modern scientific thinkers, here, in the Far West, are much more removed, in mind, from those cognate races,[1] than from the Semitic people that gave us our own Bible. These latter poets exaggerate no more than any one of us (supposing that he were a poet and not a scientific worker and registrar of hard, dry facts) would do.

But the Buddhist is nothing if not hyperbolical, and when he does magnify, he magnifies with a vengeance; take one example:—

"The Asurs, who reside under Maha Méru, are of immense size. Rahu is 76,800 miles high; 19,200 miles broad across the shoulders; his head is 14,500 miles round; his forehead is 4800 miles broad; from eye-brow to eye-brow measures 800 miles; his mouth is 3200 miles in size, and 4800 miles deep; the palm of his hand is 5600 miles in size; the joints of his fingers, 800 miles; the sole of his foot, 12,000 miles; from his elbow to the tip of his finger is 19,200 miles; and with one finger he can cover the sun or moon, so as to obscure their light."—Page 59.

And another, as follows:—

"In the forest of Himála are lions, tigers, elephants, horses, bulls, buffaloes, yaks, bears, panthers, deer, hansas, peafowl, kokilas, kinduras, golden eagles, and many other kinds of animals and birds; but the lions and kokilas are the most abundant. There are four different species or castes of lions, called trina, kála, pándu, and késara. The first is dove-coloured, and eats grass. The third is like a brown bull, and eats flesh. The késara lion, which also eats flesh, has its mouth, tail, and the soles of its feet of a red colour, like a waggon laden with red dye. From the top of the head proceed three lines, two of which turn towards the sides, and the third runs along the centre of the back and tail. The neck is covered with a mane, like a rough mantle worth a thousand pieces of gold. The rest of the body is white, like a piece of pure lime. When he issues forth from his golden cave, and ascends a rock, he places his paws towards the

---

[1] Of course, I refer to those Aryans who adopted the Buddhist tenets.

cast, breathes through his nostrils with a noise like the thunder, shakes himself like a young calf at its gambols, that he may free his body from the dust, and then roars out amain. His voice may be heard for the space of three yojanas around. All the sentient beings that hear it, whether they be apods, bipeds, or quadrupeds, become alarmed, and hasten to their separate places of retreat. He can leap upwards in a straight line, 4 or 8 isubus, each of 140 cubits; upon level ground he can leap 15 or 20 isubus, from a rock 60 or 80. When the kokila begins to sing all the beasts of the forest are beside themselves. The deer does not finish the portion of grass it has taken into its mouth, but remains listening. The tiger that is pursuing the deer remains at once perfectly still, like a painted statue, its uplifted foot not put down, and the foot on the ground not uplifted. The deer thus pursued forgets its terror. The wing of the flying bird remains expanded in the air, and the fin of fish becomes motionless."—Page 17.

Among the Legends of Gótama Buddha, there is one which has interested me very much, for it exactly corresponds with the popular ideas of the sudden, miraculous creation of living forms:—

"After eating the fruit, the sage gave the stone to Gandamba, and directed him to set it in the ground near the same spot; and in like manner, after washing his mouth, he told Ananda to throw the water upon the kernel that had just been set. In a moment the earth clove, a sprout appeared, and a tree arose, with five principal stems and many thousand smaller branches, overshadowing the city. It was 300 cubits in circumference, was laden with blossoms and the richest fruit, and, because set by Gandamba, was called by his name."—*Manual of Buddhism*, p. 306.

This creation-feat is scarcely greater than the one supposed to take place in the case of the first creation of every tree and every animal, by those who are unversed in Biology and who interpret, literally, the account of the Creation given in the first chapter of Genesis.

But some of those who have a little knowledge, even of Biology, have very misty notions of the origin of living creatures.

One such suddenly startled the writer by a hurried utterance of his cherished creed, which was as follows:—"I believe that GOD created *Wellingtonia gigantea*, 400 feet high, in a moment."

This speaker, whose power of mental deglutition was so great, was one of our own calling, and, therefore, had undergone a biological training; surely we may hope that better days are in store for us, when no educated man will be in danger of falling into such a deep pit of *belief* as this.

But Man, the only reasoning being we know anything of, can be as unreasonable in his unbelief as in his belief.

In Mr Mallock's charming little *Lucretius*,[1] there are many things a Darwinian longs to quote, but one verse, with the translator's prefatory remarks, may be given :—

"The chance to which our world owes itself, needed infinite atoms for its production, infinite trials, and infinite failures, before the present combination of things arose.

"'For blindly, blindly, and without design,
Did these first atoms their first meetings try;
No ordering thought was there, no will divine
To guide them; but through infinite times gone by,
Tossed and tormented they essayed to join,
And clashed through the void space tempestuously,
Until at last that certain whirl began,
Which slowly formed the earth and heaven, and man.'"

—Page 93.

The reader is also requested to look at the curious, abortive, unreasonable Darwinism of the chapter "On the Origin of Life and Species."—Section iv., pp. 45–50.

Those who care for defunct theories of Creation may find one as good as the rest, but more amusing, in *The Birds of Aristophanes*.[2] I can only find space for the beginning of this part of the Drama :—

"Before the creation of Ether and Light,
Chaos and Night together were plight,
In the dungeon of Erebus foully bedight,
Nor Ocean, or Air, or Substance was there,
Or solid, or rare, or figure, or form,
But horrible Tartarus ruled in the storm."—Page 30.

---

[1] William Blackwood & Sons, Edinburgh and London. 1878.
[2] Sir John Hookham Frere's Translation. Cambridge, 1883.

LECTURE II.

PROTOTHERIA (MONOTREMATA).

The lowest, or *teatless mammals*, still linger in the Australian region, in the form of the spiny Ant-eater, or *Echidna*, and the great Water-Mole (*Platypus*, or *Ornithorhynchus*). A few fossils have been found in this same region, and described by Sir Richard Owen as

FIG. 2.—Young of *Ornithorhynchus paradoxus*, one half the natural size.

belonging to a larger species of *Echidna* than any of the three known existing kinds.

But in the northern world (*Arctogæa*) no remains of any kind of Monotreme—*Duckbill*, or *Echidna*—have rewarded the labour of the palæontologist. Yet fossil marsupials have long been known in this hemisphere; although, as Professor O. C. Marsh suggests, *Microlestes* the oldest of these—the jaws of which are found at the bottom of the Secondary rocks—may belong to some simple primordial mammal, and not to a proper Marsupial. If this be so, we may have in those most precious, but also most puzzling, remains, the evidence of the existence of Prototheria (first beasts) with true teeth.

The living forms of these Prototheria are either toothless, as the *Echidna*, or have only peculiar horny plates, as a succedaneum to true teeth, as in the *Ornithorhynchus*.

That, at present, we are in the dark as to the early existence of forms of such surpassing interest to the Darwinian, is no proof that they have not had their day, and their reign, or dynasty, in this part of the world, long before the higher, or even the Marsupial, mammals appeared. For negative evidence is literally no evidence at all in matters of this kind. A few years ago we had no evidence of the existence, in the geological strata, of birds with true teeth, nor of birds with a long chain of caudal vertebræ, and distinct metacarpals. Now, however, two new additional orders of the bird class are known to have existed in the Secondary epoch, one as far below the other as the higher of the two is below ordinary flying birds; the newest of these kinds possessed teeth, and the lowest—the *Archæopteryx*—had the metacarpals, or bones of the hand, distinct, and the joints of its tail developed and elongated, whereas in existing birds they are aborted.

All attempts at keeping back the tide of modern biology are but imitations of the labours of Sisyphus;

people who do this vain thing are emulating the fame of the renowned Mrs Partington, endeavouring with her mop to keep back the Atlantic waves.

Whatever the famous Triassic jaw of *Microlestes* may turn out to be, it must not be left unnoticed that its owner co-existed with a fish (*Ceratodus*), that still co-exists with Monotremes and Marsupials in the Australian region. The living *Ceratodus* is a waif or stray from a nearly lost order of fishes—the *Dipnoi*, or double-breathers—fishes that possess both gills and lungs, and thus enjoy both aquatic and aërial respiration. Now such fishes as this Australian *Ceratodus*, as the African *Protopterus*, and as the American *Lepidosiren*, may have co-existed with the Ganoids, or fishes with enamelled armour, through much of the Primary period. Could this be proved, we should have a capital generalised stock, existing long before the period of the Trias, from which to derive all the lung-breathing forms, whether of Amphibia, Reptiles, Birds, or Mammalia. Indeed, Professor Huxley is very bold in this matter, and suggests that *Ceratodus* is a special gift of Providence, kept for these latter days, to rebuke and convince the gainsayers of the truth of Darwinism.

I cannot go into details; we walk here by faith as well as by sight. Imagine some low, ancient, simple form of fish, that did, by metamorphosis, become a creature as high as the *Ceratodus*. And that you may be able to see this hypothetical fish ascend, during its active lifetime, very far above its first condition, put before your eyes the actual transformation of the Common Frog, with the metamorphosis of which every one is familiar. Having performed this mental feat, you will have brought up from the "vasty deep" a type that, in its larval state, would, probably, be intermediate between the Lancelet (*Amphioxus*), and the Tadpole, or larva of the Frog. Now the Lancelet has no brain, no skull, slight dimensions, and scarcely any sense-organs; it is, in reality, a sort of half-way creature between a larval Sea-squid (Ascidian), and the lowest of the Vertebrata. But the Tadpole, or larval Frog, represents a low ancient kind of sucking fish (a sort of Lamprey); it has a brain, a skull, two sorts of gills, and soon shoots up into a musical, agile, air-breathing Frog.

Once more let your imaginary forces work, and feign one of these ancient double-breathing fish (*Dipnoi*), formed by transformation from that supposed low type, and you have a stock which will grow you further suckers for your life-tree.

Such a form or type, richly charged with morphological force, might transform again and again—undergo, under the stimulus of necessity, further metamorphoses. For having both outer and inner gills, and a sacculated air-bladder, acting as a rudimentary lung, it might, under the compelling force of threatening surroundings, suddenly blossom out into one of the root-types of the higher organic forms.

I am purposely forgetting, for the time, the slow accretion of minute variations, taking place through countless ages of time, and am considering sudden, *per saltum*, transformations. Whenever and wherever it became necessary that higher tracts of the drying surface of the earth should be peopled with semi-terrestrial and terrestrial forms, then I suppose these leaps of life to have taken place. The morphological force—the indwelling spirit of protoplasm—actually did perform these wonders; thus we have still living in abundance, reptiles that crawl upon the earth, mammals that march or gallop over it, and fowls that fly in the open firmament of heaven.

I do not, of course, forget that the few existing Dipnoi—the *Ceratodus* and his companions—are settled in their low estate, at their own height, on their own morphological platform, and that there is little likelihood of their undergoing any further metamorphosis, now. Still, with the *Axolotl*[1] staring me in the face, I cannot suppose even that to be impossible. But when I imagine double-breathing fishes undergoing metamorphosis in the olden times, I am thinking of more simple and archaic Dipnoi than even the *Ceratodus* or the *Protopterus* of the present day.

Every biologist knows that some types have persisted in a low estate with little modification, others in a low estate with much specialisation; whilst other types have

[1] That large Mexican Salamander generally continues in a low larval state throughout life, but now and then it becomes transformed, loses its gills, and becomes a member of a higher family.

risen altogether far above the level or platform of their ancestors.

With regard to the Prototheria (or first beasts) I am of opinion, that if the silent rocks of the Carboniferous epoch—the huge masses of Mountain Limestone—could speak, they would tell us of an abundance of teatless mammals (such as are now seen in the *Ornithorhynchus*) both on the dry land, and in the brooks and streams of water that then drained the land.

It is the business of the palæontologist to dig and bore the solid earth, and from thence extract a register from which he will probably learn that Duckbills and Echidnas swarmed in the ages of the Calamites and Lepidodendron.

But there is a poetical use of the words "lower parts of the earth," as well as a literal; in these living strata (in stage below stage of types in their pre-natal life), hidden from the sun, it is my work to dig.

These things are an allegory, and yet they are true; between the embryologist and the palæontologist there is a mavellous harmony—they have one heart and one way—*cor unum, via una*. If my valued fellow-labourer, with his huge ungainly instruments, the hammer and the pick-axe, is slow in bringing up his facts, I shall not wait for him, but, with my small needle and shears, I shall go on laying bare and spreading out the strata of the organism—a real microcosm, a world in a nutshell.

Many years ago the common Mole yielded me results

that suggested a greater nearness of this little delver to the great Water-Mole, the *Ornithorhynchus* of Australia, than had been imagined. Proof is not wanting, now, that some of the lower Eutheria, or high beasts, rose rapidly from the Prototheria, without utilising the Metatherian or intermediate condition. They did not wait to become Marsupials, but ran up on to the top platform before they attained the adult condition. The Metatheria—the pouch-bearers—did, and do still, utilize an intermediate morphological stage of development, but some of the Insectivora may have shot past them, and grown into the root-stocks of the existing noble beasts.

The Primary Edentata also may have shot up in a similar manner; but their culmination is very low, at its highest, as compared with the culmination of the Insectivora; yet I imagine them to have had abbreviated stages in their pre-natal transformation.

The registers we have extracted from the growing microcosm have enabled us to make these deductions. These writings may be likened to Palimpsests, written in many texts, one over the other; the writing, however, was not made by fallible scribes, but by infallible morphological law. This witness is true: and truth so attested may be followed by anyone, little fearing where it will lead him. I have spoken of the common Mole; I must speak of him again, in one of my lectures on the Insectivora. I will now speak of the *Ornithorhynchus* itself—the great Water-Mole.

The structure and development of this type is both Reptilian and Avian. Yet this is but a clumsy way of expressing it. The three groups—Reptiles, Birds, and low (Prototherian) Mammals—correspond in many important points, so much so, as to suggest a common root for all these three branches. In some things which are common to all three, in the number and relation of parts, these low mammals are more archaic than the existing reptiles, and very much more than the existing birds—not excluding the flat-breasted Ostrich and Emeu. But in the higher, winged birds, the parts that are distinct and simple in the *Ornithorhynchus* and *Lizard*, are found to be confluent and compound, and to undergo a practical metamorphosis into exquisite new structures for new functions.

The primitiveness of this low mammal is well seen in its shoulder-girdle; its skull, as yet only partially worked out by me, shows characters of the same sort, much more remarkably. The name given to this low order—Monotremes—suggests in one word, that which is most striking in these types, namely, that their renal and reproductive organs are constructed in a similar manner to those of a reptile or bird. In this respect, even the common Mole is a high and noble creature in comparison with the *Ornithorhynchus*, or the *Echidna*.

The limbs of the Monotremes are normally pentadactyle (or five-toed), but excessively specialised, in each case, in relation to the habits of the creature. The hip-girdle carries a pair of *epi-pubic* (or so-called Marsupial) bones as in the Marsupials, in which they lie *above* or *within* the pouch, so that their use is not very apparent, for the "pocket" swings outside them. But their morphology is evident, for the pelvis of the Salamander,

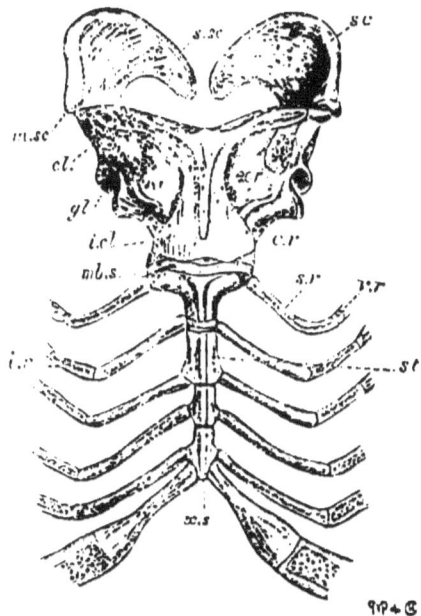

Fig. 3.—Shoulder-girdle and sternum (breast-bone) of *Ornithorhynchus paradoxus*, two-thirds natural size. *s.sc*, supra-scapular region of scapula (shoulder-blade, *sc*); *m.sc*, middle scapular region; *cl*, clavicle; *i.cl*, inter-clavicle, *cr*, coracoid bone; *e.cr*, epicoracoid; *gl*, glenoid cavity, the head of upper arm-bone (humerus); *mb.s*, manubrium (top or handle) of sternum; *st*, sternum; *x.s*, xiphoid end of sternum; *v.r*, lower part of vertebral rib; *i.r*, intermediate rib (as in *Lizards*); *s.r*, sternal rib.

and also, indeed, that of the Skate, shows similar outgrowths of the pubic region of the girdle.

But it is in the front cincture, or shoulder-girdle, that we see the most remarkable signs of ancientness in the Prototheria. The scapula in them does not give off a small coracoidal snag or beak (or imperfect coracoid), as in Man and most mammals, but the lower or ventral part of the girdle is continued downwards to articulate with the sternum (or breast-bone) as a large and perfect coracoid. Here we have a condition like that which is seen in Amphibia and Sauropsida (Reptiles and Birds); on the other hand, the abortive development of the coracoid, and the freedom of the shoulder-girdle from the sternum, is a true diagnostic of a high Mammalian.

The scapula (or blade-bone) in the Prototheria is very primitive in its form, being falcate, or scythe-shaped, and having a very low spine; the coracoid is continued from it to the sternum as a large flat bone, and the fore part of the crescentic base of the whole plate is ossified as a separate epicoracoid, a part well seen in the Frog and Lizard. Such a term as "epicoracoid" for the more or less distinct broad part of a coracoid is not needed in mammals generally, for in them the lower part of the arch is absent, although it reappears in the native Bat and Shrew. In birds the whole of the main coracoid bar is ossified from one bony centre. Here, again, the Monotremes have to be compared with types below the bird class.

Now it is well known that one diagnostic of the mammal is that it has only a pair of clavicles, or collar-bones, and that these are not simple parostoses or splints, but compound bones, composed of cartilage above and below, and of ossified fibrous tissue in the middle. But these low beasts, or Prototheria, have a large median clavicle (interclavicle) besides, and all the three bones are *simple*, being merely ossified membranous tracts.

So also are they in the Lizard tribe, and so were they in those huge whale-like Lizards of the Secondary epoch —the Ichthyosauri—whose triple clavicular structure is much like that of the Monotremes. Birds, as a rule, fuse these three bones together to make their merrythought or furcula, but in these there is a rudimentary pro-coracoid cartilage, fused with the tops of the forks of the merrythought, similar to the ordinary cartilaginous nodule on the upper end of the mammalian clavicle. The old tooth-bearing birds of the Chalk had their clavicles distinct, as in the Emeu, and in embryo birds generally. Here, again, to get at the root of the Monotremes, we must dig below the bird, and if we are safe in drawing any deductions whatever from our morphological observations, we are safe in saying that of a certainty the stock from which these beasts were derived lay as low down as that from which the earliest birds grew.

Moreover, there is a sort of solid primitiveness about the clavicles of the Monotremes, unlike what we see in the existing Lizard, in which they are very slender and graceful; they show that the best type for comparison is not the small modern Lizard, but the ancient reptilian giant—the Ichthyosaurus.

The Prototheria have the sternum (or breast-bone)

divided into segments corresponding with the costal cinctures, or arches of the chest; this is diagnostic of the mammal, yet it begins in certain Lizards, *e.g.*, the Chameleon. That which makes the vertebra of a mammal differ from that of the higher oviparous types is the development of the flat epiphyses or separate bony plates on its body or centrum.

Now these are nearly absent in the Prototheria. Albrecht and Huxley, however, have found them in the vertebræ of the tail in *Ornithorhynchus*. This fact, again, is very instructive—the Monotreme is feeling its way upwards to the higher platform on which we stand.

With regard to the skull, there is much of the deepest interest to the evolutionist, even in our present partial knowledge of its development. The *Ornithorhynchus* is by far the most primitive type; the *Echidna* has a huge brain for so foolish a creature, and it comes very near the Ant-eaters, proper, in many of its cranial characters. When I come to the *Edentata*, the group which contains the Ant-eaters, I shall refer to this fact again. At present I shall confine myself to the *Duckbill*. That which strikes the eye at once is the very amphibian look of the whole structure of the skull; it is like that of some strange Dipnoan or Salamandrian just undergoing transformation.

We, like our fellow-vertebrates, have at first a cartilaginous cranium that forms the foundation of the finished ivory casket which, in the adult, so safely holds our brain and sense capsules. But in us, as soon as it is formed into proper cartilage, it is a mere *basin*; in many types it is a box, just open above, having there a small membranous "fontanelle," as this weak part was called by old anatomists.

This *chondro-cranium*, or cartilaginous skull, is very massive in the *Duckbill*, and much of the sides and roof formed by this primitive cartilage ossifies, and forms part of the permanent skull, inside the familiar investing bones—frontal, parietal, temporal, &c.

This is in the hinder half, but the fore part, or beak, is still more remarkable as to its cartilaginous foundations.

The general form of the hind skull, or cranium proper, is intermediate between that of an Amphibian and a high mammal; the paired occipital condyles, or convex cartilaginous tracts, for articulation with the first joint of the neck, are large, vertical, and very much like those of a Frog. But the nerves of the skull, and their passages, are arranged as in the higher or gill-less types, and the hypoglosssal or motor nerve of the tongue is a *cranial*, and not a *spinal* nerve, as in the Frog. The arched structures of the ventral aspect of the head—the parts of the face and throat—are of great interest. The best morphologists differ in the interpretation of some of the details; whilst anatomists generally, those, namely, who have not been trained in Embryology, contrive to make the most absurd misinterpretations of these parts.

Here, if I fail, I shall miss both the *mark* and the

prize of the work of my life—the interpretation of the form of Man and of his vertebrated kindred. For, in reading off the characters of the *Ornithorhynchus*, and comparing them with those of the Amphibia, below, and of the Eutherian or high mammals, above, we are, so to speak, breaking the seals of a new scroll, on every line of which we can spell out the letters that go to form that great name—MAN.

That which strikes the morphologist as the most remarkable of all specialisations is the manner in which the mobile jaws of the lower type are exchanged for the fixed countenance of Man and the other mammals. In the more generalised fishes there is but little mobility of the lower jaw; but that part is carried furthest from the face in such forms as the Sturgeon among the ganoids, and in the osseous fishes generally, where the lower jaw is not close to the head, as in a Skate, but swings upon a large compound *pier*, that intervenes between the jaw and the skull. But in reptiles and birds, the hinder part of the cartilaginous upper jaw— the rest being in a great measure suppressed—forms a hinge-piece or pier to the inverted arch of the lower jaw.

In birds, generally, the whole upper jaw being mobile —or flexibly attached to the frontal region of the skull— the levator muscle, at the angle of the jaw, at once depresses the lower, and lifts the upper, jaw, and this is why these creatures are such skilled fly-catchers. Sir Charles Bell, in his charming work on the human hand, shows how this is done, and why the feat of fly-catching is so difficult in the dog trying to sleep in the sun, but kept awake by teasing flies.

In the *Duckbill*, as in mammals generally, the pier or hinge-piece is gone, and the *maxilla inferior* (or

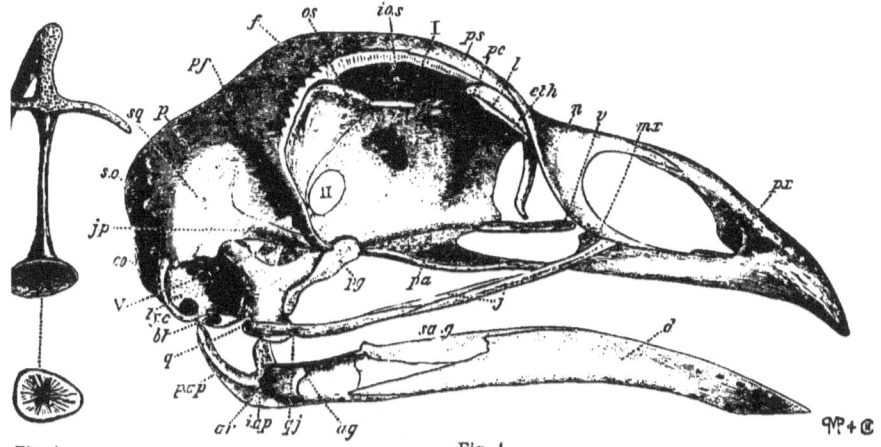

Fig. 4A.

Fig. 4.—Skull of Common Fowl (*Gallus domesticus*) one-half larger than specimen. *px*, premaxillary; *mx*, maxillary; *v*, vomer; *n*, nasal; *eth*, ethmoid; *l*, lacrymal; *pe*, perpendicular ethmoid; *ps*, presphenoid; I, olfactory nerve; *io.s*, interorbital space; *os*, orbitosphenoid; *f*, frontal; *pf*, post-frontal; *p*, parietal; *sq*, squamosal; *s.o*, superoccipital; *j.p*, jugal process of squamosal; *co*, exoccipital, V, 5th nerve; II, optic nerve; *ty.c*, tympanic cavity; *bt*, basi-temporal; *q*, quadrate; *pg*, pterygoid; *pa*, palatum; *j*, jugal; *qj*, quadrato-jugal; *p.ap*, posterior angular process; *iap*, internal angular process; *ar*, articular; *ag*, angular; *sa.g*, supra-angular; *d*, dentary.

Fig. 4A.—Auditory Columella of Fowl, magnified 6 diameters, and shown from the inside and end.

mandible) is hinged directly to the temporal bone, a solid part of the strong skull-wall. What has become of the "*os quadratum*," as the bird's jaw-pier is called? The answer is, that in the mammal there takes place a process equivalent to amputation. It is done, however, not by a chirurgeon, for it is cut out without hands.

Indeed, this is a kind of *horticultural* process, for the hinder third of the proper original lower jaw is slowly pinched off, and that hinder piece and its growing pier or quadrate cartilage are more than half starved, whilst the front two-thirds of the mandible are forced, so to speak, as if grown in a hot-bed. The starved pier or quadrate becomes the little *incus* or anvil bone of the

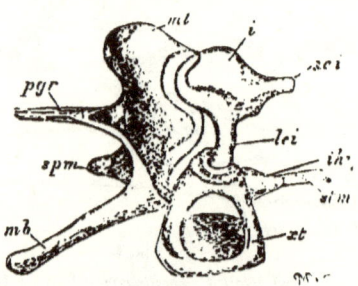

Fig. 5.—Auditory Chain of Bones from the *Middle Ear* of a New-born Pig (*Sus scrofa*), magnified 5 diameters. *ml*, malleus (hammer); *pgr*, processus gracilis of malleus; *mb*, manubrium (handle) of malleus; *spm*, short process of malleus; *i*, incus (anvil); *sc.i*, short process (crus) of incus; *lci*, long process of incus; *st*, stapes (stirrup); *stm*, stapedius muscle; *ihy*, interhyal.

ear-drum, while the hind part of the lower jaw itself, the mandible, also starved, becomes the *malleus*, or hammer; and these two ossicles are now, for the first time, added to the structures that convey the vibrations of air from the ear-drum to the labyrinth. All at once, when we look at the forcing process, whereby that which was superficial and secondary becomes the arch of the lower jaw, we see that something has been brought in for grafting purposes—a mass of cartilage which we are not familiar with in the reptile or the bird.

Looking for a slab of true hyaline cartilage large enough for Nature's purposes in these mammalian types, we travel down through Birds, Reptiles, Amphibia, Osseous Fishes, Ganoids, Skates, and ordinary Sharks, and not until we come to the extraordinary Sharks or Chimæroids do we find anything large enough. There we stop. In these waifs of an old fish fauna we do light upon what is wanted. In these fishes there is, outside

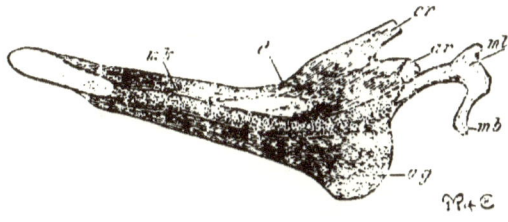

FIG. 6.—Mandible (Lower Jaw) of an Embryo Pig, 3 inches long, magnified 3½ diameters; inner view. *mk*, Meckel's cartilage; *d*, dentary bone; *cr*, coronoid process; *ar*, articular process (condyle); *ag*, angular process; *ml*, malleus; *mb*, manubrium.

the true mandible, a large slab of cartilage equal to it in size. This is a superficial (subcutaneous) band of solidified tissue, but it has no supporting bone on it. Such a supporting bone, however, is seen clearly enough in the foremost of the splints of the lower jaw (the dentary) in all fishes, except the cartilaginous kinds, and in all types above the fishes. Besides this, on the inside of the fore part of the lower jaw, in the oviparous types, there is a second splint, the splenial, and behind it a third, the coronoid. These can also be traced in the jaw of a young *Ornithorhynchus*, but they are only semi-distinct.

Then the fore part—about two-thirds of the primary cartilaginous jaw—called "Meckel's" cartilage, is ossified whilst the pinching off is taking place.

Here then let there be an end to all talk about the *simplicity* of the lower jaw in the Mammalia. After Nature has removed the hinder part, which in reptiles and birds is itself composed of three external and one internal bony centre, there still remain in the inferior maxilla, or lower jaw, of a mammal, the following elements, namely, (1) the dentary bone, with rudiments of the splenial and coronoid; (2) a large superficial cartilage, or "inferior labial;" and (3) the distal two-thirds of the primary lower jaw, or Meckel's cartilage.

Now, as to its suspension to the skull, this of course is in *front* of the old swinging point of the non-segmented jaw of the ovipara. In them the quadrate, or huge prototype of the "incus," or *anvil bone*, is attached to the skull over the tympanic cavity, and that is also the place where the incus is always found in the mammal.

In them also (the ovipara) the squamous element, or temporal scale-bone, has no cartilage on its under surface. In the bird, for instance, the "zygomatic process" is a mere snag for muscular attachment, for it has no glenoidal cavity or cartilaginous facet; there is no hinge —nothing is joined to it.[1] But in the mammal the large superficial cartilaginous tract, after serving as the matrix out of which most of the lower jaw is formed, becomes segmented into three parts at the hinge ; the lower part is the condyle, or head of the joint, the upper the glenodial facet, or shallow cup, attached to the temporal bone, and the intermediate part the *meniscus*, a sort of pad, the interarticular fibro-cartilage.

I could not find, in my young specimen of the *Duck-bill*, splints on the large rough malleus corresponding to the "angulare" and "supra-angulare," two bones that strengthen the upper or articular portion of the jaw in birds and reptiles. But I find the "angulare" in several kinds of mammals, and in the Koala, a kind of Marsupial, both of these well-known splints of the compound jaw of the ovipara are found as small separate pieces.

After long years of labour and much vacillation of mind on the matter, I am now quite satisfied that the *stapes*, or little stirrup-bone of the ear-drum, is the uppermost element of the second, or hyoid arch.

Those who have studied human anatomy know that the three little bones which are fastened as a chain across the inside of the cavity of the ear-drum or "middle ear," are called *malleus* (hammer), *incus* (anvil), and *stapes* (stirrup). The latter bone, by its base, stops up a small

---

[1] The reader who is not familiar with the skull, is referred to my figures of those of the Fowl and the Pig in the *Philosophical Transactions of the Royal Society*, 1869, plate lxxxvii. ; and *ibid*., 1874, plates xxxiv.-xxxvii.

oval window (*fenestra ovalis*) that lies between the drum cavity and the vestibular part of the labyrinth of the ear. That bone (the *stapes*) exists in Birds and Reptiles, but the other two, *as such*, do not. Also, in them, it is not a *stirrup*, but a little column (*columella*). So it is in these low mammals.

We have then, in this curious piece of morphology, no new structure, but a very new specialisation of an old one. Whatever parts grow out of, or are attached to, the columella of the ovipara, are merely processes, or at most, segments, of the "pharyngo-hyal" element of the tongue-arch, or uppermost piece of the arch.

Thus, in mammals, by a curious horticultural process, so to speak, two new elements are added to the auditory chain, namely, the incus and malleus. These parts, so modified, are diagnostic of a mammal. Why they should be correlated with mammary glands, and hair, I cannot say.

I have yet to speak of the most remarkable part of the skull of the *Duckbill;* I refer to the composition of its beak. Much as it resembles the beak of a duck, its structure is widely different, yet the superficial bones are homologous, and not altogether dissimilar; these are the premaxillaries in front, the maxillaries externally, the nasals above, and the palatines and pterygoids below.

All these bones are peculiarly thin and lathy in the young animal. They do not, as in the Duck, finish the margins of the beak; for in that bird, as in its congeners, the bones of the upper face run close to the quick that secretes the bony sheath. But the duckbilled mammal is quite unique; the whole outline of the great rostrum is formed by a large sheet of solid hyaline cartilage right and left. Over this, in front, the thin horny layer still shows the "neb" for breaking the egg-shell, quite like what is seen in Tortoises, Crocodiles, and Birds.

The extraordinary growth of true cartilage in the extended upper lip is quite similar to the growth in the lower lip of mammals generally, namely, that slab of cartilage on which the dentary bone grafts itself to form the bulk of the solid *maxilla inferior*, or lower jaw.

We have to go down, as I have just stated, amongst the lower cartilaginous fishes for similar growths of superficial cartilage in the region of the mouth. But although I am quite familiar with superficial cartilaginous structures in these fishes, it is only in the Tadpoles of the Frog and Toad, and in the adult Lamprey, that I find anything equal to what is seen in the *Ornithorhynchus*. In those pouch-gilled (marsipobranch) types, however, these parts are all separate, neat, finished tracts of cartilage, each having its place and its function as an orderly element of the front face. But in this strange remnant of a lost race of archaic mammals, the growth of cartilage is a wild leafy tract, very unlike the well-finished labials (lip-cartilages) of the types just refered to. There is, however, just one waif from the Old World, which helps us here. Of the *eight hundred* known tailless Amphibia, there are two Toads—one in Surinam, and one in the Cape region—that are tongueless, and these have their *Eustachian tubes* (or passages between the ear-drum and the throat) opening at the mid-line, as in a bird.

The Cape Toad (*Dactylethra*), has nails on his fingers and on his toes, and he is the first of the gill-bearing creatures that has taken on this specialisation—a prophecy of those exquisite supports to the finger-pulps that we see in the daintiest and most elegant of all the Vertebrates.

Now the children of this first claw-bearer are not like their parents, which themselves are the most modified of all the kindred of the Frog, but are very much like the most bizarre forms of the Ganoid fishes of the Old Red Sandstone. So much is this the case, that it is difficult to avoid the conclusion that we have in this larva, whose outline is like that of a half-opened fan, a descendant of one of those old fishes, but a metamorphosed descendant, only retaining the old family features during the time of its minority.[1]

Now this larva has a skull which differs from that of the adult Toad, into which it transforms itself, quite as

---

[1] For the transformation of the skull in the Cape Toad, and for the figures of the *larva*, see my paper in the *Philosophical Transactions of the Royal Society*, 1876, plates lvi.–lix.

much as, *or more* than, the skull of the *Ornithorhynchus* differs from that of a Man.

We may suppose the ancestors of the original teatless mammals (Prototheria) to have been something like, and not much higher than, the larva of the nailed Toad, and that these underwent an amount of transformation, during an *active out-of-door life*, equal to that undergone by the existing type. Afterwards, by little and little, such Prototheria may have improved themselves into higher and still higher types; they have had plenty of time for such changes.

ADDENDUM TO LECTURE II.

BIBLIOGRAPHY: REFERENCES TO WORKS AND PAPERS TREATING OF THE *Ornithorhynchus* AND *Echidna*.

ARMIT, Captain WILLIAM E., F.L.S., "Notes on the Presence of *Tachyglossus* and *Ornithorhynchus* in Northern and North Eastern Queensland," *Proceedings of the Linnean Society*, Zoology, vol. xiv., 1879, pp. 411-413.

BENNETT, Dr GEORGE, F.L.S., F.Z.S., &c., *Gatherings of a Naturalist in Australia*. London: John Van Voorst. 1860.

FLOWER, Prof. W. H., LL.D., F.R.S., *An Introduction to the Osteology of the Mammalia*. London: Macmillan & Co. 1876.

―――― Article "Mammalia" in the *Encyclopædia Britannica*, 9th edit., vol. xv., pp. 377-378.

HUXLEY, Prof. T. H., LL.D., Pres. R.S., *A Manual of the Anatomy of the Vertebrated Animals*, pp. 319-323. London: J. & A. Churchill. 1871.

HUXLEY, Prof., "On the Application of the Laws of Evolution to the Arrangement of the Vertebrata, and more especially of the Mammalia," *Proc. Zool. Soc.*, Dec. 14, 1880, pp. 649-662.

LÜTKEN, Dr CH. W., "A Letter to the Secretary of the Zoological Society," *Proc. Zool. Soc.*, March 4, 1884, pp. 150-152.

MURIE, Dr JAMES, F.Z.S., &c., "Remarks on the Skull of the *Echidna* from Queensland," *Proc. Linn. Soc.*, vol. xiv., Zoology 1879, pp. 413-417. See also Captain Armit's "Notes," *supra*.

OWEN, Prof. RICHARD, F.R.S., I. "On the Marsupial Pouches, Mammary Glands, and Mammary Fœtus of the *Echidna hystrix*," *Philosophical Transactions*, 1865, plates xxxix.-xli., pp. 671-686. II. "On the Ova of *Echidna hystrix*," *Philosophical Transactions*, 1880, plate xxxix., pp. 1051-1054.

PARKER, W. K., F.R.S., "On the Shoulder-girdle and Sternum," *Ray Society's Publications*, 1868, plate xviii., pp. 192-194.

At present, data are wanting to enable us to form a thoroughly clear idea of what a primary mammal, an original, ancient "Protothere," must have been like. For, at present, we are only very partially masters of what Nature has left for us to work out, namely, the structure and development of the two types that still linger on the planet. One thing we do see, however, and that is, that those two forms are far more unlike in their adult than in their embryonic (or early) condition (see Owen, I., plate xl., figs. 6-10). The beak, when undeveloped, does not differ in essentials in the two genera, namely, *Ornithorhynchus* and *Echidna* (*Tachyglossus* and *Acanthoglossus*).

In the figure of the young *Ornithorhynchus* given by Professor Owen, and in that made for me from a young specimen nearly as large as a man's fist (p. 25, fig. 2), two very important characters can be seen in the beak.

1. The first of these is that the fore part of the bill or beak arises out of a swollen basal or hind part, which ensheaths the proximal part of the free beak, exactly as in the non-flying birds (Ostrich, Emeu, Cassowary), and their nearest relatives, the Tinamous of South America.—*See* Dr P. L. Sclater, M.A., F.R.S., "On Struthious Birds," *Trans. Zool. Soc.*, vol. iv., plates lxvii[a].-lxxvii., pp. 353-364.

2. The second character is the presence of the old familiar egg-breaker on the bend of the "neb" above, just like that seen in the chick of any bird, and in the embryos of the Turtles, Tortoises, Crocodiles, and Alligators.

So that we may accuse the Duck-billed Platypus, and say—"You were hatched out of an egg and are not a proper mammal at all." But if he, like the Lamb, asserts his innocence, then we, like the Wolf, will throw the accusation backwards, and say—"Your father (ancestor) was so produced, and you are, after all, merely an oviparous creature." Probably the highest dignity this creature will ever attain to in Biology will be to be classified as an *Ovoviviparous* type, a sort of compromise between a Reptile and a Mammal. That may be true, for all animals come from eggs,—*omne animal ex ovo*,—and as Man is known to be an animal, he also once had all his potential excellences squeezed into the small space of an egg-shell—an egg which was small, indeed. We hope soon to get more light upon the development of the Prototheria. Some hunters belonging to the last and best kinds of the Eutheria are upon their track, and they must hide themselves either in the earth beneath, or in the waters under the earth, if they would escape them. It is, I think, more than probable that the original Prototherians possessed *teeth;* yet these may have been, and most probably were, of a still simpler type than those of Opossums and Kangaroos, from whose teeth we start in making an ascending survey of these organs in the Mammalia. It is not unreasonable to think that the mammals which have *degenerate* teeth, such as Sloths, Armadillos, the Aard-Vark, might serve to give us some idea of what this primitive Mammalian dentition was like. Anyhow, if they had a good mouthful of teeth, their upper and lower jaws did not resemble those of their highly modified descendants; probably they were very much like what we see in the least specialised of the living Marsupials, namely, in the Opossums of the Western World. As to their outer covering, of course they all were more or less clothed with a hairy garment, for this is correlated, always, with milk-glands; when, "in the end of the days," the last mammal appears, but appears shorn of that covering, it has to be borrowed, again, from those types in which it had not been suppressed. "Unto Adam also, and to his wife, did the Lord God make coats of skins, and clothed them." This has become, as everyone knows, a custom among the race of men, and shows, at present, no signs of becoming obsolete. Moreover, that first correlation, namely, the existence of milk-glands and a hairy covering, appears to have entered into the very soul of the creatures of this class, and to have become *psychical* as well as *physical*, for in that type, which is only, *for a while*, inferior to the angels, the fondness for this kind of outer covering is a strong and ineradicable passion. But it began, physically, as a sudden modification, with those Archaic Prototheria; yet in what forms did it appear? Hardly, at first, one would suppose, in the form of wool, rather of coarse hair mixed with spines, as in the existing *Echidna*. I strongly suspect that it struggled for a good while with the old kinds of covering, namely, scales, both bony, in the skin proper, and horny, in the cuticle. If so, the existing Edentata, of which I shall speak anon, are a much more precious legacy of time and nature than they have hitherto seemed, even to the most enthusiastic biologists. In a purely technical paper on this group, lately read before the Royal Society,[1] I have said that the fact that in the Armadillos the new husbandry, or growth of hair—the correlate of milk-glands—thrives badly on the old stony ground of Reptilian horn-covered (bony) scales, breaking out where it can among the clefts—is not more wonderful than that this same new growth of hair in the Pangolin should mat itself together, and imitate the scales of Reptiles and Fishes.

If this be true of those placental descendants of the almost oviparous Monotremes, much more, one would suppose, must it be true when we are speaking of the very first evolution of a hairy creature. Indeed, if the first creature clothed by the Creator, to speak enigmatically—after the manner of men—with the hairy skin of a beast, did, *per saltum*, gain his hairy coat in one great metamorphic leap, it would be nothing wonderful if some of his descendants should backslide a little, and, under degenerating influence, now and then show some mark or stigma of the old Reptilian nature. In the paper just quoted, speaking of the scale-covered Pangolins, I have remarked —" If the term *Reptilian* might be applied to characters seen in any placental mammal, it might to what I find in this. This creature has most remarkable correspondences with the Reptilian

---

[1] See *Proceedings of the Royal Society* for June 1884, p. 80.

group. Of course, the scaly covering is mimetic of the Lizard's scales, and is in reality made up of cemented hairs; that may pass; but not the structure of the sternum in some species, with its long 'xiphisternal horns,' as in the *Stellionidæ*, and the cartilaginous abdominal ribs, as in the Chameleons, and some other kinds."[1] In the poverty of the existing, but highly modified, Prototheria, we are glad to get any addition to our materials for work, any knowledge that may help us in our deductions. As I shall soon show, the Edentata are only a sort of Eutheria, or high kind of mammal, *quoad hoc*, in this and that point in their organisation; in other respects they have kept in a low estate, having the slow temper of some races of men, who are haters of change, however beneficial, and of whom it may be said "as their fathers did, so do they."

My task in writing of these types, after straining the eyes of my mind to see what sort of folk those mammalian forefathers were, is rendered more difficult through my being precluded the free use of technical terms. A rustic gymnast in a sack, with nothing but his homely features free, and yet having the necessity of jumping laid upon him, is not more an object of sympathy than a biologist, when robbed of his familiar terms—his special nomenclature. As the movements of the one are of necessity a series of jerks, so the thoughts of the other are too often put into language that to an easy-going, well-trained writer must seem to be spasmodic. Loosening my bonds a little, however, and taking a few technical liberties with the reader, I will endeavour to give some of the remarkable evidences to be found in the *quasi reptilian* nature of these primordial beasts. In the mid-region of the Vertebrata, especially amongst the Serpents and Lizards, we come across some very remarkable structures in the fore part of the organs of smell; these are called "Jacobson's organs." They were described by Rathke, in the Snake, under the term "nasal glands"; that term was adopted by me in my papers on the Skull of the Snake and of the Lizard.[2] In those papers the contained organ was not described, as not being in my plan,

---

[1] See my memoir on the "Shoulder-Girdle and Sternum," *Ray Society Publications*, 1868, plate xxii. fig. 13.

[2] *Philosophical Transactions*, 1878, plates xxvii.-xxxiii., and 1879, plates xxxvii.-xlv.

but the bones and cartilages that encapsule it were carefully described, and copiously illustrated. In those types we have the culmination of these organs, which have some mysterious connection with the organ of smell; the *first* or *olfactory nerves* give off fibres to them. If these organs have their *height* in these Reptiles, they have their *decline* in Man, who, however, in an early stage, possesses them, as Professor A. Kölliker's invaluable researches show, both those published at Leipzig in 1877, and those much later, at Würzburg in 1883. For an abstract of this last piece of research I am indebted to the excellent "Summary" in the *Journal of the Royal Microscopical Society*, for April 1884, pp. 201-203. The concluding sentences of this abstract are as follows:—"From the rich possession of nerves by Jacobson's organ in an eight-week old embryo, and their disappearance in older embryos, we may conclude that the organ is now in a rudimentary condition as compared with what it was in ancestral forms."

Now, on one hand, in Serpents, and Lizards, we have these organs and their related skeletal parts, both cartilaginous and bony, highly developed and persistent, and on the other, in Man, these organs are soon aborted; nor am I aware that the skeletal parts that should support them are more than feebly developed. Although we are in a deplorably agnostic condition with regard to these organs, they may be used as a measure of the height of any mammal or order of mammals, in the scale of life. In my young specimens of *Ornithorhynchus* (the size of a moderate fist, with the hair appearing), these parts and their capsules are as large as in Serpents and Lizards. In the Marsupials, Edentates, and Insectivores they are well developed in the embryo up to the time of birth, and for some time after, having considerable persistence in several at least of those kinds. They are present in all sorts of mammals, as far as research has gone, at least in the embryo; and the bones and cartilages that support them are more persistent than the organs themselves. In the Reptiles, these organs are mostly invested by bone, in the Mammals they are well encapsuled by cartilages growing backwards from the snout. In the Mammals only one pair of small bones assists in protecting the soft gland-like organ; in the Reptiles it lies on each side, as in a dish, formed by a bone, the so-called vomer (ploughshare), and it is covered in by another bone which serves as an elegant lid, not present in the mammal, the so-called "turbinal."[1] Those who are biologists, and care to go into this matter, will do well to refer to Wiedersheim's *Lehrbuch der Vergleichenden Anatomie der Wirbelthiere*, Jena, 1883. In my forthcoming paper on the Edentata there will be found a Bibliographical List of various Memoirs and Papers on these organs.

But the general reader will see that there is some big secret shut up here, and that, as far as it has been found to disclose itself, it is all unimpeachable evidence in favour of the gradual development of the higher, and even the highest, forms of animal life; those curious parts of the nasal labyrinth that have had their rise and their decline in the various Vertebrata, now coming into morphological and physiological importance, and then having a feeble and a fading growth—these facts must now be added to the enigmas of Biology. The structure of our body is full of old things as well as new; the old things have had their day, but they are abrogated, and to us practically they are "beggarly elements." But the new things are not really new; they are merely expansions and improvements, so to speak, of things as old as the hills. It is just possible that in the Vertebrata of the Primary Rocks some rudiment or other existed of every structure that has now completed its evolution in the human body. But some one,—one whose mind carries no biological ballast, is always starting up and making the demand of a sudden creation of Man. Let him learn of the great Theologian, St Augustine, that the Creator of all things is *patient, because eternal*. This is exactly what modern Biology teaches: whatever the force is that worketh all in all, it is certain that it has had, *practically*, for all purposes of adaptive variation in organisms, unlimited time. There has been plenty of time, in the gradual accretion of gentle and almost insensible modification, for an almost unlimited amount of variation; but *per saltum* changes have often occurred; that is quite certain. These outbursts, so to speak, of morphological modification during the individual life of a creature—Butterfly or Frog, for instance—are amongst the most amazing of all pheno-

---

[1] Both those paired bones were named as above by Cuvier, and both erroneously; the Science of Embryology scarcely existed in his days, and many of these things can only be interpreted by an embryologist.

mena. They remind one of the sudden and mysterious moral perfection of the antediluvian prophet Enoch, of whom it is said, that—"He being made perfect in a short time, fulfilled a long time." There is, however, another way in which the mysterious morphological energy works, so as to force, as in a hothouse, the growth of the young of certain types. This is the case in the highest sorts of birds—the "Altrices," or high-builders—that have tender nestlings. These young, sweltering in their soft nest, and almost smothered by their feathered mother, whose blood is nearly at fever heat, grow and develop at ten times the rate of the young of those birds that make poor nests on the ground; for those chicks, hatched strong and lusty, grow slowly to perfection. Yet the temperature of the blood itself is apparently equal in both cases, and in both cases affects the *temper* of the mother-bird. At that time the true maternal courage rages; at that time "a Wren will peck an Estridge;" and a Hen, the gentlest of mothers when her brood is grown, is like one possessed whilst they are young.

So we see that Nature fulfils herself in many ways; her works have not gone on from age to age in tame and cold uniformity, but in the plenitude of her morphological energy she has at sundry times, and in divers manners, burst out into new developments—delivering herself in her mighty energy of myriads of new and wondrous births. Let us imagine ourselves living in the time before the beginning of the reign of the Prototheria, and before the first feathered creature grew, when there were neither birds nor beasts, and to us it seems to be unenlivened gloom; we have in idea depopulated the planet of almost all the living forms that make it laugh and sing. The vision is like that of the prophet who—weeping over the desolations of that land which once flowed with milk and honey—says, "I beheld, and, lo, there was no man, and all the birds in the heavens were fled." Now, if for the sake of *Biology* we should be glad to repeople the earth with the parents of the Prototheria, for the sake of *Life* we should indeed be sorry to peel away the newer Mammalian faunas until we got to that old core.

To sum up these Prototherian matters, we may now look at some of the most remarkable characters in the *Duckbill* and *Echidna* that are manifestly reptilian, or *quasi-reptilian*:—

1. Jacobson's Organs are about equal in their development to

what we find in Serpents and Lizards, where they have their culmination.

2. The skull itself has a strong and thick foundation of cartilage, the ossification of which forms much of the permanent cranial box, whilst the superficial bones are flat and relatively small, as in such a reptile as the Lizard.

3. The fore face has a large vallance of solid cartilage, such as is not seen again, until we get down to the most archaic of the larvæ seen in any metamorphosing type whatever—for example, in *Dactylethra capensis*, in whose *larval skull* a similar vallance of cartilage grows copiously.

4. The lower jaw is seen, even in the adult, to be the equivalent of the fore part of a Reptile's mandible, whilst the *malleus* (hammer-bone) is manifestly the hinder part of such a mandible, and has, cemented to it, a most rudimentary ear-drum bone.

5. The anvil-bone (*incus*) has not taken on the normal Mammalian form, but is a mere flat segment of ossified cartilage ; it is a very small "quadratum" or equivalent of the *hinge-segment* of the Reptilian mandible.

6. The *stapes* (stirrup-bone) is not normal, it is merely a Reptilian *columella*, or little column, with a dilated upper end.

7. The shoulder-girdle is perfect, both in its complete moieties of ossified cartilage, and in its superadded triple clavicular plates of simple ossified membrane.

8. The vertebræ of the spine, as a rule, are devoid of the normal Mammalian bony plates that are added, fore and aft, to the body of each vertebra, to take off the shock in the quick movements of high and agile types.

9. The hip-girdle repeats the so-called "pre-pubic" bars seen in Salamanders and in some kinds of Skate.

10. The organs for the growth and maturation of the germs are, in essentials, quite like those of Reptiles and Birds, and there is, as in them, no differentiation or subdivision of the terminal outlet.

So that these creatures are just plucked out of the Reptilian group by their *skin*, with its hairy covering and its rudimentary milk-glands. Severely apply to them the rule "*Cucullus non facit Monachum*," and say—"The skin does not make the beast," and back amongst the Reptiles they would have to be driven.

After writing the above, I received from Professor Simon H. Gage, B.S., of Cornell University, Ithaca, N.Y., a most important communication on the respiration of certain fresh-water Tortoises. The information thus given covers just a page and a half, and yet it is of more value to the biologist than some bulky volumes that one could name. I shall insert it, bodily. Let the facts there disclosed be but fairly considered, and the difficulty of supposing a gradual melting down of the distinctions between the Amphibia and Reptilia will be at an end.

The lining of the *pharynx*, or upper part of the gullet, is the proper normal respiratory organ of any creature possessed of a *notochord*, or *primary spinal axis*. All the various specialisations that may be found in *Ascidians* (Sea-squids), *Amphioxus* (the Lancelet), and in all the *Vertebrata*, are of secondary importance to the embryologist. The peculiar structure and functions of the pharynx described by Professor Gage may be due to *degradation* or *relapse*, but if so, it only proves *that the aquatic was once* the mode of respiration in the stock from which these Tortoises sprang.

"Pharyngeal Respiration in the Soft-Shelled Turtle (*Aspidonectes spinifer*). By Simon H. Gage of Ithaca, N.Y.

"During the last twenty-five years the mechanism of respiration in the Chelonia has been investigated with considerable thoroughness, both in this country and in Europe; and at present the Chelonian form of respiration is considered to be comparable with that of the mammal rather than with that of the frog, as formerly supposed. While, however, the mechanism of respiration has been quite fully investigated, there has been, so far as I am aware, but one who has considered the organs of respiration in the different groups of turtles.

"Professor Agassiz, in Part II. of the *Contributions to North American Zoology* (p. 284), states that the lung capacity of the soft-shelled turtle is far less in proportion to its body-weight than is that of the land turtles. He also states, in considering this fact, that the skin on the ventral side of the body, from its rich network of blood-vessels acts as a respiratory organ. He further states that in the pharynx are many fringe-like processes which resemble the inner gills of tadpoles, and probably have the same function, although no mention is made as to the method of their use.

"In 1878, while watching a soft-shelled turtle from Cayuga Lake, confined in a glass aquarium, it was observed that the throat and the floor of the mouth became alternately swollen and collapsed, while the turtle was completely immersed in the water. The appearance was very much like the respiratory movements of a frog in the air. As no air escaped from the turtle, the bulging of the throat and mouth must be caused by filling the mouth and pharynx with water, and expelling it, or the air must be forced into the mouth from the lungs and then forced back into the lungs, as is done sometimes by men when swimming under water.

"In order to determine whether or not water was taken into the mouth and expelled, the bottom of the aquarium was covered with fine sea sand, and the observations were made when the animal was resting quietly on the sand.

"At the beginning of the movement, the mouth would slightly open, and its floor would swell out, the swelling passing steadily onward to the throat. After a moment of quiet, the swelling would disappear in the inverse order of its appearance.

"During the disappearance of the swelling of the throat and mouth, the sand, for a considerable distance in front of the animal's head, would be swept aside as by a rapid stream. The movement of the sand, without the escape of air, seemed to prove conclusively that the mouth and throat were alternately filled with water and emptied.

"These pharyngeal respirations, as they may be called, were very regular, occurring ten or fifteen times a minute. My observations were verified by Professor Wilder and several of our laboratory students.

"While, therefore, the investigations of Agassiz showed that in the pharynx of the soft-shelled Turtle were organs apparently suitable for aquatic respiration, the observations here recorded of the rhythmical bathing of these organs with fresh water seem to make the evidence complete, that a true aquatic is combined with an aërial respiration.

"It is hoped that during the coming year investigations may be completed which shall determine the exact amount of oxygen consumed in this pharyngeal respiration, and the structure of this unusual respiratory organ in the soft-shelled Turtle."—From the *Proceedings of the American Association for the Advancement of Science*, vol. xxxii., Minneapolis Meeting, August, 1883.

LECTURE III.

ON THE MARSUPIALS, OR POUCHED ANIMALS (METATHERIA).

IN their structure generally, the Marsupials are intermediate between the Monotremes and the placental or nobler forms of mammals (Eutheria). With regard to the development of their embryo, these types are far below the Eutheria, while they are most probably above the Prototheria; there, however, we are at a standstill for want of materials.

In their early development these pouched mammals come very close to the higher oviparous animals (Reptiles and Birds), but we have good reasons for asserting that they are less specialised, or more archaic than the existing members of the great group of Sauropsida (Reptiles and Birds in one group). All these kinds—the Reptiles, Birds, and Marsupials—may have been on the same level once,—may have arisen from the same old *Amphibian* stock—but the embryo and its wrappings, in these mammals, is in some important respects less developed than in the other two groups.

It is a very instructive fact that the tail-less Amphibia

—Frogs and Toads—should show a tendency to become members of the *Amniota*, the higher gill-less types which develop an *amnion* and an *allantois*, two of the three membranes that enclothe the embryo.

No other explanation of the curious pouch developed in front of the lower part of the great intestine of the Frog, when it is passing into an air-breathing type, can be given, than that it is a rudiment of the allantois. Yet this is all; there is no amnion correlated with it, and only in Reptiles, Birds, and Mammals have we all three of the fœtal membranes or bags—yolk-sac, amnion, and allantois. But there is every reason to suppose that other temporary gill-bearers, forms on a level with the existing Amphibia, did, during transformation, develop the rudiment of the amnion as well as of the allantois. Such Proto-amniota, under the quasi-magical influence of new surroundings, may, very probably, have grown into forms, of which some kinds had a huge development of all the three bags, while in others they all appeared, but were all arrested.

In none of the Marsupials or of the higher mammals is there more than a slight development of the yolk-sac; none have any bulk of food-yolk, so that the embryos of both these groups are dependent upon some other source of food. In all cases the mother supplies this pabulum, but whilst in Birds and Reptiles the food is prepared in the ovary and its duct, in the Metatheria or

Marsupials that source of supply exists for a very short time, and then the mammary glands furnish the table. So also do they in the high types of mammals, but not until after a very long interval. In these the food-yolk is extremely small in quantity, yet the embryo is not supplied with milk when that ceases, but receives its nourishment through the instrumentality of an enormous allantois, united with the highly vascular walls of the dilated oviduct.

But even in the Eutheria there is a great variation

FIG. 7.—*Uterine* embryo of Virginian Opossum (*Didelphys virginiana*), magnified 6 diameters.

as to the time at which this last and newest mode of feeding an embryo comes into play, some kinds having tender, and others precocious, young.

The observations of Professor Chapman of America show that these bags are all present in the Kangaroo, but that they are all small and arrested, so that neither the allantois, as in the higher mammals, nor the yolk-sac, as in some Sharks, grow into more than a temporary union with the oviduct, or uterus, so as to derive nourishment from its walls. Hence the small size of the young of these creatures at birth, the food-yolk being so soon exhausted, and no other pre-natal supply being at hand.

Two newly-born young of the large Kangaroo (*Macropus major*) sent to me by Dr Bennett of Sydney, were not so large as new-born Rats, *i.e.*, they were about an inch long. Yet these small Kangaroos, whose

FIG. 8.—New-born young from the pouch of Kangaroo (*Macropus Major*), magnified 2⅜ diameters.

parents are the size of Sheep, like the sedentary Oyster "attend at ease moist nutriment," being attached to the teats, and there abiding. The embryo of the Marsupial is comparable in some degree to that of many fishes, in which the food-yolk is soon exhausted, and the embryo develops rapidly, the cartilage and bone appearing very early. Such is the case in the *Lepidosteus*, or Bony Garpike, and in the Sturgeon.

In these young Kangaroos (an inch long) the ossification of the skull is much advanced, as I found to my sorrow when sections were made of them by the *microtome*; in young, three-fourths ripe, of the Virginian Opossum, the size of the larva of a Blue-fly (fig. 7, p. 61), the development was also very advanced, and the cartilage quite solid. There is a considerable amount of contractile or muscular tissue in the teat of the mother, and the structure of the throat of the young is such that the syringing action of the walls of the duct does not choke it, the larynx passing up above the soft palate. There is thus, as in the Cetacea, a direct passage of air from the external nostrils to the glottis (or opening of the windpipe), and fluid can pass right and left of the breathing tube, with no danger of choking.

This temporary modification of the young Marsupial to conditions which, for a time, endanger its breathing apparatus, is well worthy of notice, albeit it is but one amid ten thousand instances of the ready response of the organism to the influences by which it is surrounded. Indeed, the earth and the inhabitants thereof, may in some sort be looked at, collectively, as a great, infinitely complex organism, and the working of the whole, as if it were one body with many members, may justly be called earth-life.

In this group—the Marsupialia—the general intelligence is low, the brain has but few convolutions, and the great familiar bridge, over which nerve-impressions travel right and left—the corpus callosum of our brain—is thin and feeble in these types. At first sight these Metatheria appear to be a very neat group, a people quiet and secure, having no business with any other tribes, but living in their own zoological seclusion. But, as in all similar cases, this is only apparent; the hedge set about them, and all that they have, is as unsubstantial as a dream. Yet they are a feeble folk, and their structure, habits, and distribution in space and in time are all congruent to this view of the pouched tribes.

The noble beasts, like the nobler tribes of men, are "mighty hunters," and they have driven the feebler Marsupial tribes before them. Hence it is that in these days, "Wallace's line" bounds them on the north, in the Eastern world; while in the Western continent, only one genus (Didelphys, or the Opossums) lingers amongst the Eutherian types, and one or two species have found their way over that great western *world-link*, the Isthmus of Panama, yet the real home of that genus is in the southern, and not properly in the northern half of the American Continent at all.

But the loss of so many of these low types, in this

"conquest of the Canaanites" is only the partial working of a general law, in which Nature is always doing for every great fauna what the farmer does who seeks to improve the breed of his cattle.

From year to year, as you may perhaps know, the sheep are brought "under the hands of him that telleth them," and he, guiding his hand wittingly, judges with quick motion, which are fittest to be next year's mothers, and which are to be appointed for slaughter.

His wisdom and intelligence are great, but how little, as compared with what his great Earth-mother—the farmer of farmers—has shown ever since the green earth was first stocked! Of this huge farm, with its great unenclosed tracts of pasturage, the forms least able to bear changes of condition die out first. But there are various ways in which such changes necessarily affect living creatures, and one of these is caused by the frequent immigrations made by the herbivorous tribes as the pastures become bare, and by the carnivorous tribes who follow them for plunder. Change of feeding-ground means also change of climate, more or less, to hotter or colder, to wetter or drier. Instinct, as we all know, is only an imperfect guide, and the animal tribes have to learn. Their tact is not always inborn, or always accurate; as an instance, I will mention one familiar to me from childhood. When our upland sheep—used to close, quick fences—are removed to the fen-districts, where the fields are enclosed by straight canals, there are

E

always some losses by drowning, until the sheep learn the meaning of those dark waters.

The earth has often had her shaking fits and her attacks of colic, and then the living creatures suffer with their mother; those that escape are the strongest or the most cunning; those that can "rough" it in new homes, or that are deftest in escaping from danger. Nature has, unconsciously, adopted this rough method of culling out her weaker tribes—appointing them for slaughter—and of saving the best for the increase of the flock. These suggestions relate to the incoming of the Eutheria, of which I must treat soon; if Nature had not dispossessed the Metatheria, and placed nobler beasts in their room, we ourselves—the Eutheria of the Eutheria, the noblest of the noble—should have had no existence.

I now pass from that old occupation, Husbandry, to this new work, Embryology; and if the reader will give me a little attention, I will show him reason for believing that the Marsupial group arose from similar low forms to those that gave origin to us and the nobler beasts, some of which, indeed, may be transformed Marsupials; and that the line of demarcation between the nobler and less noble types does not form a perfect fence.

In the study of nature, as every one knows, that seems to be the most bewitching part in which each particular observer is working; the skeletal framework takes the precedence with most of us. There are many excellent diagnostic marks in the skeleton of the

Marsupials, some well known, some less familiar to anatomists, but none of these are absolutely wanting in the forms above, nor are there any that cannot be traced to the forms below.

Yet, fagoted together in the Metatheria, these diagnostics serve the purposes of the classifier, and are, indeed, in their combination, remarkably distinctive of the group.

Firstly, those parts of the skeleton which are popularly supposed to be so important in relation to the pouch—the marsupial or pre-pubic bones—these are no new thing, but, as we have seen, are equally large in the Monotremes, which have the pouch rudimentary; they also exist as pre-pubic cartilages in Salamandrians and Skates; and they reappear, as rudiments, in the Eutheria. The pouch itself—a superficial structure, a mere apron or fold of the skin—reappears, as a rudiment, in the embryo of the Flying-cat of the Phillippines, a sort of primordial Bat, not quite out of the border of the Insectivora. Such a pouch holds the eggs of Pipe-fishes, but it is the male which possesses it; and some Frogs have such a pouch, but it is on their back. The shoulder-girdle of the Marsupials is quite like that of the highest kind of mammals, in which the clavicles are well developed; there is no interclavicle between them, and they have a pro-coracoid rudiment at each end, a thin cartilaginous pad. The *acromion* (shoulder-point) is well-developed, and the *coracoid* (Crow's beak process) is a mere spur, as in the Eutheria generally. So also the vertebræ, the ribs, and the sternum; these are quite normal, having taken on the characters we are familiar with in the higher kinds.

The limbs are very instructive. The front pair are very similar to what we find in Insectivora and Rodents, and are not much modified, but the hind pair are very much specialised in those that can take long leaps—as the Kangaroos. The Opossums of America, and the Phalangers of Australia, have the hallux (great toe) short and opposable, as in the Dormouse, and in the Quadrumana (Apes and Monkeys), generally. From such a foot as that there is every transitional form to that of the Kangaroo, where the hallux or first toe is gone, the second and third extremely long and delicate and evidently useless, whilst the fourth is very large, and the fifth moderate.

One peculiarity of these pouched animals is seen in their dentition. Several years ago, Professor Flower showed that only the third false grinder (pre-molar) has a predecessor. This milk tooth is like a true molar; the tooth pattern is simple, quite unlike that of the high Herbivora. The Rodents, which are lower, are extremely variable in this respect, and the dentition, in some of them, assumes a high condition, with extreme specialisation.

As the metropolis of a country is the most instructive as well as the most important of its towns, so the skull takes to itself that which is best and noblest in the organs of the body. It has cost me but a moderate amount of attention to this group (which I am now taking up in earnest) to find in the skull ten good diagnostic characters. These are easy to follow by anyone at all familiar with the skulls of the higher mammals. I must therefore beg the attention of the human anatomist, who will at once see how curious and suggestive these deviations are from what is normal in our own species, and indeed I may say also in most of the higher kinds. Nevertheless, these deviations are not confined to the Marsupials, but are to some extent seen in many of the lower kinds of Eutheria, and are in themselves nothing abnormal at all, but only so in relation to, and comparison with, the standard we have set up, by making our own structure the measure of all others. Indeed, these peculiarities are so many stepping-stones between us and our highest Mammalian relatives and the forms that lie below; we are not so isolated as we have supposed ourselves to be.

The things which strike the eye in the examination of the Marsupial types of skull are as follows:—

1. In the basal region of the nose there are several pairs of splint bones, belonging to the vomerine series, besides the large middle vomer or ploughshare bone, like that which sheaths the base of the partition of the nose-labyrinth in us.

2. That strong floor, the hard palate, which in us and our congeners divides the cavities of the nose from

the upper region of the mouth, and which is formed by a special ingrowth of bone from the maxillaries and palatines, is large and imperfect where the four plates

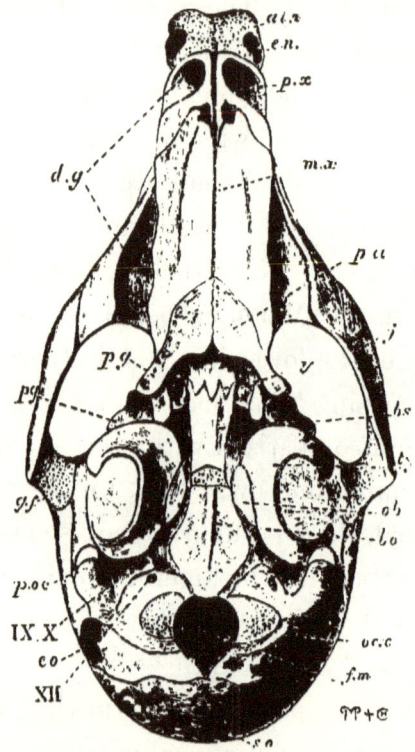

Fig. 9.—Skull of an embryo Pig, 3 inches long, lower view, magnified 3½ diameters. *al.n*, alæ nasi (snout cartilage); *e.n*, external nostrils; *px.*, premaxillary; *mx*, maxillary; *d.g*, dental groove (for tooth-pulps); *pa*, palatine; *py*, pterygoid; *v*, vomer; *e.pg*, external pterygoid plate; *bs*, basi-sphenoid; *g.f*, glenoid fossa (hinge cartilage for lower jaw); *ty*, tympanic bone; *ob*, os-bullæ (additional tympanic bone); *bo*, basi-occipital; *p.oc*, paroccipital process; *oc.c*, occipital condyle (hinge for first vertebra or *atlas*); *co*, exoccipital; *so*, superoccipital; *f.m*, foramen magnum (great opening); IX. X. XII., holes for the exit of the 9th, 10th, and 12th cranial nerves.

should meet. It is less imperfect in the embryo than afterwards, but the bone-cells are very thinly scattered in the early state, and during growth become aggregated laterally, so as to leave large fenestræ, or windows, towards the mid-line.

3. The temporal bone (squamosal) is hollow above the hinge of the lower jaw, and this large air-cell communicates with an extensive series of similar empty spaces that arise primarily in the mastoid region, or back part of the organ of hearing.

4. The bony ring of the ear passage (external auditory meatus and drum cavity), does not form all the cavity, but a hollow shell of bone from the hinder wing of the sphenoid (ala magna or alisphenoid) applies itself in front and within, so as to form what is called an "auditory bulla." Yet this bulla or bleb-like shell of bone, a part that we do not possess, does not correspond with that of the Cat, whose meatus-skeleton has a large shell-like inner tympanic bone added to the usual annulus, or ring-bone.

5. The internal carotid artery does not enter the skull, as in us, between the basal beam on one hand, and the side-wings and petrous bones on the other, but burrows through the basal beam, each branch appearing in the seat of the turkish saddle, "sella turcica," instead of burrowing the "petrosal," and then passing through a "foramen lacerum," or ragged interspace between that bone and the lateral parts of the skull.

6. The malleus, or hammer-bone of the middle ear, has a very large processus gracilis, which, be-

sides growing well forward into the Glaserian fissure, also sends a large sickle of bone in front of, and within, the bony tympanic ring; thus, that bone has two ancillary pieces, helping it to wall in the drum cavity.

7. The innermost bone of the middle-ear chain is not always stirrup-shaped, it is often a mere "columella," or rod, with an oval dilated plate above, where it fits into the oval window (fenestra ovalis), as in the Monotremes and Ovipara.

8. The angle of the lower jaw-bone is greatly incurved, and often has on its upper and inner face a hollow fossa.

9. The hyoid bone is not simply U-shaped, but is dilated into a wedge-shaped plate; it has, however, the usual cornua (horns or processes), that are indeed the skeletal parts of the hyoid, or second arch, and of the rudimentary third, or first gill arch.

10. But the most interesting and instructive of all the characters is one which at first sight would seem to be a very small matter, but is indeed full of instruction, namely, that the optic or visual nerve does not pass from the brain to the eye through a special hole in that part of the skull, but through a large chink in the walls—the common outlet for all the nerves of the orbit.

These are the ten good, useful, well-marked diagnostic characters, which I promised to show in the skull of a Metatherian animal, or intermediate beast.

It now has to be shown, in the interests of biology, that these are not present in the Marsupials as absolute characters, and that they are not formed here, either for the first or the last time. In other words, they are mere specialised modes of structure, quite familiar to the student of the oviparous types (either lower or higher), and they do not die out suddenly, but reappear in the Eutherian types that are far above them in zoological height.

Every element of the skeleton of any of these classes, however inconspicuous, is a link in a very long chain, and often to the morphologist a golden link, very beautiful and valuable, suggesting to him origins and ends that would have been unintelligible but for some such small points of bone or nodules of cartilage. On the other hand, the proud conservative, who would isolate himself upon his human throne, must not think that we are removing biological landmarks, we are merely showing him that they never existed.

I will now take these ten diagnostics one by one, and look at them in their rise and in their progress.

But let me not be misunderstood; their rise cannot be seen by us in any actual progenitors of these Metatheria, nor their progress in types that have arisen from them. That is absolutely impossible in the nature of the case. The lower types are mere survivals of races more or less on a level with the various supposed stages through which a Metatherian must have passed to attain to its present height. And the types that have gone beyond the Metatheria are not the children or descendants of any existing Metatheria. The parents of the lowest Eutheria have been quietly inhumed for many an age, and there is indeed no reason to suppose that they all utilised the Metatherian or pouch-bearing stage at all; but it is probable that they shot past the Marsupials in an embryonic stage. Many things seen in the Marsupials seem to suggest this; they have formed for themselves a sort of "bye-path meadow," far out of the line of the great highway of life.

Letting imagination fold her wings for a while, we will now look at a very few dry anatomical facts.

Character 1.—In Bony Fishes, but not in the more archaic Ganoids, there is a median vomer. In the lower Ganoids, *e.g.*, the Sturgeon and Paddle-fish, there is a crop of such bones, with a tendency to a quincuncial arrangement, that is, with a middle series, and a row, right and left, opposite the interspaces of the single row.

Then, in the Amphibia, and in Snakes, Lizards, and Crocodiles amongst the Reptiles, there is a pair of vomerine bones, but only in the Chelonians (Turtles, and Tortoises) is there a single median bone.

In Birds also, certain groups show a single bone, as in rapacious Birds, Fowls, Geese, &c., in other kinds there are two, either for a time, as in Finches, or permanently as in Woodpeckers. But the vomer and the vomerine series of bones have behind them, under the main skull, another series, namely, the "parasphenoid" and its divisions. This series, as I shall afterwards show, appears in the Mammalia, and the arrangement is always as one, two, or three—a recollection, so to speak, of the primary pattern of median and sub-median bones in the lowest Ganoid Fishes. In the mammals, generally, during the embryo stage, there are five vomers, but in Marsupials there may be ten.

Character 2.—The bony palate is deficient in the Hedgehog and other low Eutheria, and is very limited in the lower Rodents. Such a specialisation of the cheek and palate bones is only rudimentary in the Amphibia, in Serpents, and in Lizards; in the larger Chelonia (Turtles) it is very considerable, while in the Crocodile, as in some of the lowest Eutheria, *e.g.*, Ant-eaters, it attains its utmost development. In birds, which, more than any other group, lie away from this line of descent, this structure is very slightly developed.

Character 3.—This character, the hollowness of the squamous part of the temporal bone, is very marked in the lower Eutheria, such as Edentates and Insectivora. In the tailed Amphibia there is no drum-cavity; in the tail-less kind, where it is generally present, I never saw this cavity enlarged by extension of the air-cell into the neighbouring bones. Nor in Serpents or Lizards is there any excavation of the bones in this part; the former have no drum cavity, most of the latter have.

But in the lesser Tortoises the tympanic cavity is made quite large by the hollowing out of both the quadrate and the squamosal, whilst in Crocodiles and Birds the whole hind skull, at any rate, is one system of air-galleries, all communicating with the cavity of the drum.

In us, whatever kind of ear-drums our very first parents may have possessed, there are no cells of this kind except in the "mastoid process," the thick mass below the labyrinth, which we feel as a lump behind our ears.

Character 4.—Whilst writing these notes, I have for the first time found this fourth character in a mammal above the Marsupials, namely, in an Insectivore from Zanzibar (*Rhynchocyon*), a creature full of inconsistencies, but a treasure to the Darwinian. To him who can wait, the whirligig of time brings its rewards as well as its revenges. That mixed type (of which Dr. Dobson was the kind donor) has come to me for the establishment of my faith in development.

Another equally valued friend, Professor Burt Wilder, of Cornell University, U.S., amongst other treasures, sends me unborn embryos of the Virginian Opossum, and now, after years of patient longing, I can compare the development of this type of skull with that of the Crocodile and the Bird.

The process of cartilage that grows out on each side of the second part of the skull-base, the hinder sphenoidal region, which forms the rudiment of the excavated part that enlarges the drum cavity on its inner side, is developed in the same manner in all the three types. There is much difference in detail, but the mode of growth, as well as the primordial condition, is alike in all these—Crocodile, Bird, and Opossum.

Nature, who has framed strange fellows in her time, must have gone to the limits of her power in growing a Crocodile, a Nightingale, and an Opossum out of germs as like to each other as the right hand is to the left. These forms had, undoubtedly, an hereditary *something* in them that determined each along its own diverging line. The angle of divergence is very acute, but ultimately the distance has become wide enough.

Character 5.—This character, the entrance of the internal carotid artery through the substance of the basi-sphenoid, seems at first sight to be a very little matter. It is, however, a character correlated with a lesser brain and a lower intelligence than we find in the better sorts of the Eutheria. I see an approach to this state of things in the little Ant-eater, a very ancient kind of creature, of a very non-intelligent sort.

No doubt, if anyone would carefully give himself up to the investigation of the modes of arterial supply, he would find that there is a most orderly series of changes in the development and distribution of these vessels, and of all the arteries in the body. But the interest attached to those which go to form the "circle of Willis" in Man is the greatest, as it is one of the principal means of supply of fresh blood to the brain.

Anyhow, as the brain widens and grows, in the ascent of the types, these arteries get further from the midline at this their entrance into the skull. In Birds and Reptiles the pituitary body drops into a hole,—does not lie on a saddle as in us,—and in them the internal carotids run up through the open space, close beneath the pituitary bag.

Thus, we see that the Marsupials are intermediate between the Sauropsida and the Eutheria; they have an imperfect seat to their "sella turcica" (turkish saddle, as it is generally called in Man), and the internal carotids pass inwards, as in the oviparous types, but rather further apart.

Character 6.—The peculiar form of the malleus (hammer) seen in many Marsupials. Its large size, and the sickle-shaped tympanic fork of the processus gracilis, is seen in low Eutheria, *e.g.*, the Mole, for a time.

The growth of periosteal or superficial bone is an attempt to form the usual splints of an ordinary mandible in the ovipara. Sometimes the three plates—external articular, angular, and supra-angular—are quite distinct. These are formed just as the cartilaginous rod is being pinched in or starved off, so that the fore part of the jaw becomes segmented from the hind part, does all the mandibular functions, and leaves the hinder, or upper end, to partial starvation, in order that it may be small enough to form the special malleus.

But the wrapping and binding of the cartilaginous lower jaw in bony splints—three on the fore part, and three on the hinder, proximal, or upper part—is as old as the Ganoid Fishes of the Old Red Sandstone, who formed their mandibles in that way; and the living Sauropsida —all known Birds and Reptiles—to this day form their mandibles in the same manner; it is a new thing when the hinge-part of the lower jaw is starved, so that it may shrink into the little auditory malleus.

Character 7.—This also is a Mammalian modification —a new specialisation of an old structure—when the innermost bone of the ear-chain grows so as to resemble a stirrup. It does so because of a peculiar branch of the "common carotid artery," which passes in the early embryo close to the opening (fenestra ovalis) of the vestibule, to press its way to, and unite with, the "inferior maxillary artery." In its passage it trespasses on the ground belonging to the topmost segment of the arch of the tongue (pharyngo-hyal), and a compromise is made by the cartilage hardening round the little artery. But in several of the Marsupials this does not take place; nor in several of the lower Eutheria, *e.g.*, some of the Edentata. The Monotremes, also, the Sauropsida, and the Amphibia, have this element imperforate, and thus it has received the name of the "auditory columella."

In the Skate this element is a large triangular cartilage, on which swing both the first and second arches of the face; hence it has been termed the "hyo-mandibular," for it carries the hyoid or tongue arch, and the mandible also.

Character 8.—The incurved form of the lower jaw at its hind angle is not confined to the Marsupials; it is also seen to a less degree in various kinds of lower Eutheria. In the types beneath, it simply represents the large development of the dentary bone; in many of these it forms a "trough," and grows freely on the inner side of "Meckel's cartilage."

Character 9.—The basi-hyal plate, or skeleton of the back of the tongue, is seen in the higher mammals, where this plate is enlarged in relation to the larynx. In Turtles and Crocodiles this plate is flat and wide; in the Frog also it becomes a broad sheet, like an apron, the narrowed bands of the arch being the strings. But in fishes there is a chain of segments at this part, the basi-hyobranchial series of *key-stones* of the series of gill-arches. It is thus short in the air-breathing types, because of the suppression, in them, of the gill-arches.

Character 10.—This is a "small wire," but it will bear to have a large amount of induction hung upon it. I know of no case in the existing Vertebrata below mammals, where the optic or visual nerve is protected by a special bar behind it from the plate that forms the base of the skull at this part ("orbito-sphenoid")

Above the Marsupials I have, as yet, only found two cases,—the Common Shrew and the Rhynchocyon,—in which it is not thus protected, and these small beasts are among the lowest of the Eutheria.

The teaching of all these details is manifestly a doctrine of development. Every new fact (and new facts are pouring in day by day) certainly makes the old theory of creation more and more untenable. The long time during which these pouched beasts have existed, their intermediate position between the Prototheria and Eutheria, their present isolated position, and their distribution only in territories where archaic forms most abound—all these things look in the same direction, and tell us the same tale.

There is something very remarkable in the manner in which this group, composed of forms so nearly related to each other, has been broken up into families, *mimetic*, or imitative, so to speak, of the Orders of the Eutheria. This appears like an attempt on their part to make the best of themselves on their low level. Nature says—"The Eutheria have I loved, but the Metatheria have I hated;" and yet these latter, also, have attained to much increase, and to a rich variety of life.

But their date is nearly out. Almost everywhere, in the Northern World, they are extinct; and wherever the chosen people (the Eutheria) come, there these low types will of necessity die out. The noble races show no mercy to the ignoble; when Nature elected that

F

the Eutheria should increase, and multiply, and fill the land, then she practically culled out, and appointed for slaughter, these poor silly pouch-bearers, the Metatheria.[1]

ADDENDA TO LECTURE III.

The numerous forms, still happily existing, of the Marsupials, have given rise to so large an amount of biological activity, that it is impossible here to mention more than one or two quarters where treasures of this kind are to be found. Of course, the late Mr Gould's valuable works are to be referred to, and also the important papers, with illustrations, that from time to time appear in the publications of the Zoological Society.

The reader interested in their anatomy may refer to the works, given in the first list, of Professors Huxley and Flower, and to my own paper on the "Shoulder-Girdle and Sternum," where those parts of the marsupial skeleton are described. An excellent and well-illustrated memoir on the muscles of the limbs and other parts of the anatomy of this group, by Dr. Cunningham, will be found in the *Challenger Reports*, vol. v.

Remains of the huge extinct forms of those types of the Australian region long ago found their historian in Sir Richard Owen, and invaluable papers, freely illustrated, describing those huge creatures, will be found in the *Philosophical Transactions*; for twenty-five years these have been gradually coming to the light.

These are but a fraction or division of his work. The huge extinct relatives of the Goose, the Rail, and the Emeu—from another part of the Australian region, New Zealand—have also been revealed

---

[1] From all that we can gather of the history of the "stocking" of the earth by mammals, the low-brained kinds have always given way to those with more developed brains. The utmost specialisation of peripheral parts cannot compensate for a small brain with low intelligence; the cunning, and its twin-faculty *invention*, of dogs and foxes, give the dog-like Marsupial (*Thylacinus*) not the least chance; and a few goats and ponies would soon drive out or starve whole herds of kangaroos. One man with a pocket-knife will do more than another with a whole chest of tools.

to us in a similar series of memoirs in the *Zoological Transactions*. These, and many others come flocking to my mind whilst I write. Having begun my own biological studies by the help of some of his earlier works, I take pleasure in making mention of these noble monuments of the labour of a long life.

The great men who did so much for palæontology half a century ago, of whom Owen is almost the only survivor, have now their rivals in the present generation of American geologists, who are daily giving us new and pleasant surprises.

But, whilst one man will "cut Colossus out of a rock, another will carve a head in a cherry-stone."

Great as is the value to be set upon the work of our palæontological fathers, the work of the rising biologists is of still greater value; and that even with regard to the past history of these Metatherian types. Palæontology is good, but Embryology is better, for if all Sir Richard Owen's giants could be made to live again—an exceeding great army—they would tell us less of the origin of the Marsupials than we should gain by the knowledge of the development of a single germ of any one living type.

Mr J. J. Fletcher, B.A., B.Sc., has lately sent me two of his papers, the beginnings of his researches into the anatomy of the internal organs of the Australian Marsupials. These have been published in the *Proceedings of the Linnean Society of New South Wales*, Nov. 30, 1881, part i. (Introductory), pp. 796-811, and Nov. 31, 1883, part ii. pp. 6-11.[1]

There have also come across the Atlantic, lately, two noteworthy memoirs; the first, "On the Embryo of the Kangaroo," is by Dr. H. C. Chapman (*Proceedings of the Philadelphia Academy*, 1881, part iii. p. 469), and the second, *Observations upon the Fœtal Membranes of the Opossum and other Marsupials*, is by Henry F. Osborn, Sc.D.

The two papers just mentioned might, literally, be folded up and packed inside a nut-shell, and yet, if I am not greatly mistaken, they let in more light upon the incoming of both the Metatheria and the Eutheria than anything that has gone before.

[1] On the same page (11) there is a paper by Mr C. W. de Vis, M.A., "On the Remains of an Extinct Marsupial, a new Type," called by him *Sthenomerus charon*.

Of course, only the biological reader of such communications can value them properly, as he only can thoroughly understand their meaning and their bearings; and yet the patient and thoughtful general reader may come at the gist of the matter. But first of all he must bind up all his old misconceptions into a bundle and burn them, and come into the Temple of Nature with child-like simplicity of mind, and not like that "praying, synagogue-frequenting beau," who walked up to the Holy Place straight as a ram-rod—stiff with pride and prejudice.

We know that in the oldest oak-tree living there has been no discontinuity of vital action since it first germinated, and that its germ, with the closely-packed cotyledons, were not created, but grew.

These first, or *larval*, leaves, the cotyledons, had their day, and did their day's work like honest labourers, so also did the first crop of normal leaves, and the second crop, and so on, year after year, all doing the work of their generation. This increasing family of leaves budded and grew, and stopped in their growth and died; but they all helped to make that forest-king.

As long as we think of the oak, merely, we can run back for some length of supposed time, along the line of our imagined oak-tree's ancestors. But if we were to go some distance back—very far back, no doubt—we should find ourselves lost, for our oak would be the English kind no longer; we should have reached the point where all the species of oak would meet in one generalised type. But the oak family has its relatives, and these would, far enough back, all become undistinguishable. Along such a descending road we should never find rest until we had reached the common, most generalised, protoplasmic mother-stuff, whose descendants, on our return journey, would turn out to be every green thing, every plant or herb or tree that the earth has borne, or is still bearing. This thrice-ancient mother of all the plants is the same as she who was the fruitful mother of all sentient creatures or animals. From her sprang the fishes of the sea, the fowls of heaven, and the beasts of the earth. If we could follow the pedigree of every living, moving creature, it would be traceable back to that common protoplasmic mass. "Wisdom" saw the green corn, the rose, and the oak; and also birds, and cattle, and men, in their first beginnings. Solomon says—and he, the wisest of the sons of men, should know—that She rejoiced in, and that her delights were with, those living creatures, whose evolution was, as yet, so far off.

I think that Solomon, who rejoiced so greatly in the living creatures of his own country, would have been less enthusiastic over the lower and less beautiful types of Australia.

The aboriginal human inhabitants of that strange country—Australia—the home of that grey, cheerless vegetation, and of those lowly unintelligent quadrupeds, have never shown any signs of mental evolution; at their best they have risen but little above the dignity of a "connecting link."

Palestine, most probably, and England, we know, did once possess a Flora and Fauna the counterpart of that now existing in Australia, but we have no evidence with regard to that human type, in its archaic state, which in the fulness of its evolution, long afterwards, took to Poetry and Biology. There is an unusually thick mist over that matter, a darkness that is felt, for its effect is very disappointing and distressing to the modern type of mind. There is no doubt that Darwin has roused our curiosity about these things to an almost morbid degree; but let us look in the direction where the mist is lifting a little. Nature has set the Marsupials in the very midst of the higher tribes of the Vertebrata, and especially of the Mammalia: they are the mid-beasts. Sir R. Owen long ago got some light upon the peculiar position of the Marsupials, in relation to the *truly viviparous* Mammals; Dr. Chapman has lately helped the matter considerably, and Professor Osborn has thrown new and most welcome light upon it; and he will not rest until the problem of the development of these types is fairly solved. One of those poetical emotional Easterns rises into the most joyous raptures over the early development of a man, of himself indeed; but he turns his thoughts all to poetical and devotional purposes. He had the stuff of a biologist in him, but lacked the proper training. Let us look at the beginning of this mystery of development in a lower type. If we take a very low kind of fish, say the river Lamprey, we shall find that it lays a countless number of small eggs that are hatched in the water, and the fry are like so many small black worms. Very little care is taken by the Lamprey, and by most of the common fishes, of their spawn and fry, but in some cases, as in the Pipe Fish (*Syngnathus*), the eggs undergo their development in pockets or pouches in the abdominal region of the *male*. Some river fishes of the *Silurus* kind (*Arius*, &c.) lay a few large eggs—as large as a small cherry—and as these fishes have in them the beginnings of family affection, and yet are *nomads*, travelling about a good deal, up and down mountain streams, the male pockets those precious globes in his mouth, whilst he and his mate wander about, laying them down whilst they rest.

The Amphibia (Newts and Frogs and their kindred) lay small eggs, as in the Lamprey, with but little food-yolk, but a number of curious family arrangements are to be seen in these groups.

The Common Frog lays multitudes of eggs, and each of these at first is covered with a tenacious jelly, which, swelling with the water, expands to the size, and has the appearance, of a white currant; thus an egg-cup-full of eggs soon grows into a mass that would fill a quart measure, each white globe having a blackish egg inside, the size of a mustard-seed. The father and his wife hang about these, their future progeny, until they are hatched, but do not seem to understand the meaning of them, or that they are being eaten, by hundreds, by the Duck, whose voice might seem to indicate a Ranine descent. As for the Obstetric Frog, the male coils the slimy eggs round the thighs of the female, and she swims about with them attached to her, a family arrangement not unlike what we see in the Shrimp and her kindred.

But these tail-less metamorphosing Amphibians, that in their transformation give us the promise, or anticipation, of much that finds its culmination in our own body, have also some very remarkable ways of showing their social and parental affection and care.

This extremely ancient family holds its own, and keeps its place, in the presence of all the newer and nobler types of Vertebrata. Their fecundity enables them to lose a large proportion of their progeny without diminution of their actual numbers, year by year, a considerable percentage slipping through the fingers, and out of the very mouth of fate. Moreover, we yearly see our common kinds taking hold of the forelock of time, and getting their water-bred brood out of the way before the drought can kill them.

Some of the *foreign races* have very curious habits.

"In *Notodelphis oripara* the eggs are transported (by the male?) into a peculiar dorsal pouch of the skin of the female, which has an anterior opening, but is continued backwards into a pair of diverticula.

The eggs are very large, and in this pouch, which they enormously distend, they undergo their development. A more or less similar pouch is found in *Nototrema marsupiatum*.[1]

"In the Surinam Toad (*Pipa dorsigera*), the eggs are placed by the male on the back of the female. A peculiar pocket of skin becomes developed round each egg, the open end of which is covered by a gelatinous operculum. The larvæ are hatched, and actually undergo their metamorphosis, in these pockets. The female during this period lives in water. *Pipa americana* (if specifically distinct from *P. dorsigera*) presents nearly the same peculiarities. The female of the Tree Frog of Ceylon (*Polypedates reticulatus*) carries the eggs attached to the abdomen.

"*Rhinoderma Darwinii* behaves like some of the Siluroid fishes in that the male carries the eggs during their development in an enormously-developed laryngeal pouch.

"Some Anura do not lay their eggs in water. *Chiromantis guineensis* attaches them to the leaves of trees; and *Cystignathus mystacius* lays them in holes near ponds, which may become filled with water after heavy rains.

The eggs of *Hylodes martinicensis* are laid under dead leaves in moist situations."—Balfour's *Embryology*, vol. ii. p. 100.

Now, if *Bees* "teach the order of a peopled kingdom," so also do Frogs and Toads suggest to us all sorts of skilful ways of keeping a family from want and danger.

But in no case, yet, have we in this survey come across any type that preserves its developing progeny *within itself*, nourishing and cherishing it, not merely as a nursing mother does her child, but as the same mother does while that child is still invisible.

That method is nature's most motherly invention. Does she give any prophecy that shows any beginnings of this exquisitely gentle forethought in the cold tribes of the waters?

A good time before our era Aristotle discovered that certain sharks have this habit. "*Mustelus levis* (the *smooth hound*, as the sailors call it), which is one of those in which development takes place within the uterus, presents a remarkable peculiarity in that the vascular surface of the yolk-sack becomes raised into a number of folds, which fit into corresponding depressions in the vascular walls

[1] Here, certainly, the pouch opens behind.

of the uterus. The yolk-sack becomes in this way firmly attached to the walls of the uterus, and the two together form a kind of placenta. A similar placenta is found in *Carcharias* (white and blue sharks)."— Balfour, *op. cit.*, p. 54.

Now, having looked at the smooth and blue and white Sharks, we seek the shore again, and are, once more, in Australia, among Kangaroos and Phalangers; then we voyage across to South America; afterwards call at the Southern States of the North, and lastly, in imagination, get home once more; for the Hedgehog, the Mole, and the Shrew will require our attention. Indeed, whilst writing about the Marsupials, it seemed to me that "Thrice and once, the Hedgepig whined."

A few words more about the Metatheria; after them the Edentata must be spoken of; and then we can look at the prickly infants of the Hedge-hog.

The Marsupials have all three embryonic membranes—*yolk-sac*, *amnion*, and *allantois*—sub-equally developed, but not equal to what is seen in Reptiles and Birds, which have the yolk-sac as large as in the Sharks and Skates but have, besides, the two other membranes, which have no existence in fishes, as such, although the fishy Frog gets a rudiment of the *allantois* during transformation. But in Reptiles and Birds the egg-coverings, membranous mostly, or both membranous and calcareous, intervene between the developing embryo and its inner enfoldings, and the walls of the egg-duct (*uterus*); this membrane may burst in the act of *laying*, and thus the young be born alive, as in the Vipers. In the Eutheria, or higher Mammals, all these membranes are developed, but there is scarcely any food-yolk, much less, indeed, relatively than in Fishes and Amphibia that lay very small eggs.

In the Eutheria the non-vascular amnion and the highly-vascular allantois are highly developed, and it is the latter membrane, and not the yolk-sac, as in the Shark, which forms that wonderful and most perfect union and inter-communion with the equally vascular lining of the enclosing organ.

But the Marsupials have, as I have said, a moderately large *yolk-sac*, as large as, and even larger than, the other two membranes, and this forms, for a time, a commercial union with the walls of the uterus in a manner like that of the Shark, but this union is only temporary.

The *allantois* does the same thing, beginning to form its mysterious union with the lining walls of the uterus, as in the Eutherian embryo at the same stage.

But this strange union of the Shark method with the method of the high Mammal is soon broken off, for, whilst yet a most minute creature—in the Virginian Opossum half an inch, in the Kangaroo an inch, long—the little creature is born and placed by the mother in the pouch, and on the nipple.

During its pre-natal life the embryo develops very rapidly, and the form, whilst so extremely small, is well finished,—a new-born kitten is scarcely better made,—and it manages, in its new nursery, to hold on, tarrying nature's leisure, until its bones become well knit, and its sinews strong.

The single genus (*Didelphys*, or the Opossums), still found in the Western World, might yet, through long periods, and suffering from mutability, gently and slowly become parental to any number or kind of new Insectivores; but we shall not live to see such changes.

The Eastern Marsupials, however, have, without transforming into Insectivores, done some grand feats within their own circle, pre-figuring in some degree several of the Orders of the higher sorts of Mammalia. Their fossils show that some kinds were scarcely less than the existing Elephants, whilst some living species have diminished into a sort of pouch-bearing Mouse.

Large or small, dead or living, they all have a kind of Reptilian 'smack' about them; and if the Mammalia had gone no higher than the highest of the Marsupials, the Cosmos would have had no terrestrial student, and the Author of all would have had no intelligent praise ascending to Him from the inhabitants of the green earth.

LECTURE IV.

EDENTATA.

LIKE the pouch-bearing tribes, the Edentata are fast disappearing, and like them they are found only, or almost only, in the southern regions; in this case, the greater proportion are natives, not of the eastern, but the western, division of the "Notogœa" or southern hemisphere.

There, in South America, the common herbivorous tribes are scarcely represented at all; a small spiny pig. the *Peccary*, a few deer, and an ancient form with a short proboscis—the Tapir—are the largest native cattle; the latter is found also in the eastern tropics. In the South African region, two genera of Edentata still exist, namely, the Pangolin (*Manis*) and the Aard-vark (*Orycteropus*); the former is still found also in India and Ceylon.

These types, which are either toothless or have imperfect teeth, and not any in the front part of the upper face, teach a lesson which every Darwinian must learn if he would escape from very grave mistakes; I refer to the extreme specialisation often to be seen in the low types of any Order or Class.

In the latest paper on the Vertebrata for which I am responsible, namely, that on Marsipobranch Fishes (Lamprey and Hag), just published in the *Philosophical Transactions*, I have ventured to state that the lowest known fish—the Hag—is as much specialised for its own kind of life, as man himself is, for his. In some of my former Lectures I showed in detail the extreme modification, for special purposes, seen in the structure of Serpents.

Now Serpents are always reckoned to be the lowest of the Reptilian class; they are indeed degraded forms, going on their belly, because they have lost their legs, and eating dust with their meat because of their helplessness. But if nature has taken from them with one hand, she has freely imparted her gifts to them with the other, and their spine and skull beggar all other forms of animal mechanism.

There has been plenty of time for both loss and gain in the ascent of the various existing forms from their simple, metamorphosing, hypothetical ancestors; and in the countless ages during which this agonistical wrestling for life has gone on, changes have taken place that we are only slowly beginning to conceive of. Tens of thousands of types have fainted and failed, and have dropped out of the race and out of the conflict; of these, here and there, a word may be read in the leaves of the old stony book. Others, however, wrestled until they were almost exhausted; nevertheless, they held on, faint yet wrestling still; and the blessing of continued existence was pronounced upon them. Now, the Serpents, as it seems to me, are among the worthiest of all those that had power with their antagonists; they fought upon their stumps when their limbs were almost gone, and when the roots of their limbs were lost, then they still wrestled and fought.

If life had not been a struggle with the angel of death, our existing Edentata might have been left by nature, as defenceless as the first human types. They were, however, her dull, but dear, children; and she helped them as she did not help us and our ancestors. She did not change her plans for them; but she clothed some with a thick heavy fell of hair, and others she harnessed with coats of mail.

What science wants to embody forth is the primordial root-form of all the nobler creatures now existing; all those, I mean, that *begin life* by breathing air, all the forms above the Amphibia (Frogs and Toads, Newts and Salamanders). Indeed, we need the conception of a still lower root-form than the Tadpole or the larval Newt, one which might possibly have been the parental form of all the existing Vertebrata.

Let us imagine such a primordial form, and then we can argue that the lowest existing Vertebrates, the Hag and the Lamprey, are the descendants of types that have had as much leisure for improvement as the forefathers of the noblest kinds. All these types, during countless ages, have had a talent to occupy with; this talent, so to speak, was the innate inherited morphological force.

As to ascent in the scale of vertebrate life, some have made their one talent ten, others two; those, however, which did not improve themselves, even in the midst of untoward surroundings, had to pass into the darkness of extinction. Even such a type as the Lamprey shows a great stride beyond the form whose *imago* is still seen in the temporary Ammocœte (Sand-pride, or larval Lamprey).

The Singing-Bird had, probably, at one time, an ancestor but few removes from an Ammocœte. But the former comes of an ambitious stock; and now, in the end of the world, whilst the home of the Lamprey is still in the sand and the mud, the surroundings of the Bird are the trees, and the sky, and the laughing sunshine.

Myriads of types, unready and conservative, have passed out of being; that which they had, but did not improve, has been taken from them, whilst others, by steady improvement of what they had, have had more and more life given to them. But not always by slow, steady increment "during long ages" has the advance been made. Nature does, now and then, make amazing leaps, certain types taking on sudden metamorphosis, and, in the fraction of a life time, the low is transformed into the high.

These beautiful creatures—the Edentata—have been

richly endowed with specialising force; nature working by, or rather *in* them, has produced marvels of organic growth and architecture, but this has been mainly for purposes of safety. Here, however, we meet with what is not seldom seen in human life, where the finest forms are not always the fittest for permanency. Especially is this the case where the body has been fed at the expense of the brain.

Why such a form as the Glyptodon should have failed to keep his ground is a great mystery; nature seems to have built him, as Rome was built,—for *eternity.*

His exquisite little relative—the Chlamydophorus—

FIG. 10.—Embryo (one third ripe) of 9-banded Armadillo (*Tatusia hybrida*), a little above natural size.

scarcely larger than a Mole, has continued as yet to run out of danger, safe in his littleness; and many other kinds of low-brained Mammals have, so to speak, the power to make themselves practically invisible.

The living Armadillos are not all "minims of nature," although none of them can compete with their lost relative, the Glyptodon. These forms, with their bony cases, horn-covered, are not such strange members of the Mammalian class, to my mind, as the Pangolins, with their merely horny, imbricated scales. At first sight, we seem to have in these old-world Edentata types that have relapsed into a reptilian condition; they are, however, merely mimetic of the scaly tribes. The Armadillos come nearest to the Reptiles, such as the Blind-Worm, the Tortoise, and the Crocodile, whose

 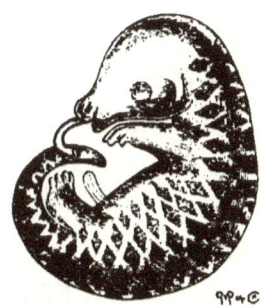

Fig. 11.—Embryo (one-third ripe) of a Pangolin (*Manis* sp.), reduced to ¾ natural size.

bony scales are encased in horn, but none of them fail to develop at every available place, that glory of the Mammal—the hair. But the Pangolin is tiled over with patches of cemented hair. In the early embryo, lozenge-shaped tracts of skin are seen all over its body, with lines of thinner cuticle between.

Under the microscope, sections of these thicker tracts show that they are composed of fine hairs cemented together by a copious growth of epidermic cells; here and there larger hairs are seen, but these fail to reach the surface, turning again towards the inside, like nails driven into wood that is too hard for their points.

Nothing could be more distressing and disgusting to the highest type that wears a covering of hair than this matting together of these fine, delicate filaments that in their rich fulness crown the head of Beauty: when it does occur it is abnormal, and is relegated to pathology. But here, in the Pangolin, we see a perfect coat of mail formed by the imbrication of these large tracts of well-cemented hair-plates. Such an armour would, if the sun were overhead, "scald with safety;" but the Pangolin is a lover of the shade, and, although not very intelligent, is much too wise to allow himself to be burnt under his own roof-tiles.

The other Old World Edentate—the Aard-vark (*Orycteropus*)—is not covered with armour, but with a thick coat of coarse hair. He is not truly edentate, but has teeth similar to those of the Armadillo. Similar, but not the same, for his sub-cylindrical rootless teeth have many vertical pulps instead of one. A section of these shows a compacted mass of hexagonal prisms; so that his teeth might be said to be a sort of ivory whalebone.

Most biologists think that these are a degradation of proper mammalian teeth. Long ago, Professor Owen, in his magnificent *Odontography*, showed that the various sorts of Cartilaginous fishes have teeth whose section shows them to be similarly composed. This is a fact worth noticing.

But this earth-pig—it is a flattering title, he is far below a pig in the scale of life—discourses eloquently to us on the subject of extinction. If ever there was a lonely creature this is one; if ever there was a generalised type this is one; he has no near relative, and his nature is that of a very peculiar kind of Armadillo, *without armour*, and with a dash of the true Ant-eater in him; but quite unique, all the while, having characters that he only possesses.

There may have been a time when species and genera of Aard-varks were as plentiful as blackberries, but that time is past; it *cannot* be that he is the single descendant of a single line of mammalian life, that never broke out into forks or branches. Taking the Edentata as a whole—Ant-eaters, Pangolins, Armadillos, Sloths—(but leaving out the Aard-vark), I am strongly of opinion that they shot up suddenly, so to speak, from the old prototherian stock.

In the next group I have to treat of, the Insectivora, I shall show reason to believe that these lingered, in many cases for a time, in the metatherian territory; some of them seem to be improved Marsupials (Opossums).

Not so the Edentata; these, I fancy, most of them, went through their metatherian stages in the dark; they did not utilise embryonic characters that correspond with those utilised in the permanent forms of the Marsupial animals. Indeed, that eastern monotrematous "Ant-eater," the *Echidna*, appears to me to be a highly modified Protothere with the external modifications of an Ant-eater; arrested inwardly, and not transformed throughout, but retaining the old quasi-reptilian characters in the deep-seated parts of his organisation. Afterwards, in summing up this subject, I shall show that we ourselves, by the force that works in us, long before we become independent and conscious creatures, hurry through a series of changes parallel with conditions that are utilised and permanent in stage after stage of lower types.

The Edentata lead nowhere; they are the end and consummation of all that came before them in this particular line of mammalian life; they have culminated in forms of low intelligence. If we could trace their past history, and see them rise by transformation of the old early forms of mammalian life, their genealogical tree would be fairly typified by a spreading vine of low stature, or an old well-weathered umbrella pine.

Yet, in their birth and their nativity, they, perhaps, promised as well as the Insectivora; they must have stood, so to speak, on the same level or platform—once upon a time. But nature, caring for their mere personal safety, with all this care, failed in the nurture and development of their brain substance; this correlation of a strong body with a weak head, is not uncommon in the highest mammal we are familiar with.

The evolution of protective characters is seen in this group, I think, better than in any other.

The Sloth is not only well wrapped up so as to be "warm o' nights;" he is stained of a greenish colour, so as to be more like the mosses and lichens that grow on the patriarchal trees of the primæval forest than any proper living creature. Now this greenish tint in his coarse, spongy hair, as Mr Sorby has shown, is due to the growth of an alga in its interstices. Here, again, what would be loathsome in us, is made by nature to be a most wholesome and safe thing for this, the slowest of all the Mammalians. Nevertheless, this overdoing of protection becomes a real danger to these beasts; they take less care of themselves, and have less and less power to escape danger.

On the other hand, the most naked and defenceless of all the tribes of the earth has become the lord and governor of all the tribes; in his case, the protoplasm which has been saved from the skin has gone to the service of the brain.

Coming to particulars as to the structure of the Edentata, I may remark that the scapula has a very large coracoidal tract, as compared with that of the higher mammals; this is seen especially in the Sloth and the Ant-eater. The Pangolin, I find, has at times, eight cervical vertebræ; Sloths may have as many as nine, and as few as six. Here we encounter ancient characters, such as undoubtedly existed in prototherian forms, for all normal mammals, from a Man to a Giraffe, have seven joints in the neck.

As to the limbs, the greatest loss of digits has taken place in the *Unau*, or Two-toed Sloth. In most of the Edentata the modification has been rather in the form than in the number of the digits. Of course, the skull is the most important part of the framework of these types, as of all others; this is one of the best corners of the field for the worker, for here is the deepest and richest soil.

As compared with Marsupials, these types show a skull of a higher order, on the whole; yet, whilst none of the known Metatheria have lost their teeth, their skull seems to be more archaic than that of the Ant-eaters and Pangolins. In the Armadillos, Sloths, and Aard-vark, the dentition is degraded to such an extent as to put these forms far out of the line of the Mammalia, generally, whilst the teeth are entirely suppressed in the Ant-eaters and Pangolins. This entire absence of teeth, common to two of the groups, and to the Protothcrian *Echidna*, gives a peculiarly feeble appearance to the face. Also, as in birds, this part is very much drawn out (prognathous); in Turtles and Tortoises, where the face is short and steep, the loss of teeth does not seem to modify the head so much as in Birds and Ant-eaters.

The youngest embryo of an Edentate worked out by me was that of a Pangolin (*Manis*), and the likeness of the skull in this stage (the embryo being only the size of the little finger) to the skull of a bird is very great.

In the Crocodile's embryo, when the teeth are just appearing, but have influenced the form of the jaws very little, and in the embryo of the bird, at about the middle of incubation, we have excellent subjects for comparison with the early stage of the Pangolin.

The Pangolin, albeit one of the Eutheria, is in reality but little advanced above the Marsupial platform. The simple form of its facial bones, not hypertrophied, to make room for the teeth, is excellently fitted for special morphological observation; we have the essential, without the non-essential, in these parts. On the whole, this skull is very much like that of an Insectivore at the same stage, but the interpretation is made easier in this case by the more primitive form and condition of the investing or superficial bones.

If the skull had been arrested at the stage I refer to, the difficulty would have been to separate it, as a type, from the skulls of the Sauropsida (Birds and Reptiles) generally. But, gradually, after the middle of the intra-uterine period, changes take place in the structure of the various parts, that ultimately amount to what is, practically, a thorough metamorphosis.

The changes I am now about to discuss are so great

that the embryo, before they take place, may be considered as bearing an *essentially larval relation* to the permanent and highly modified form. No one need be startled at the term *larval*, nor at what will seem, perhaps, still more daringly speculative, viz., that this quasi-larval stage has been inherited, and that it was not always *quasi*-larval merely. Such forms may have lived, and most probably did live, an open life, with a slow, real metamorphosis; there is no great difficulty in supposing, also, that the waters brought them forth abundantly, and that they were equal to their aquatic surroundings; not hesitating to develop gills, when, as yet, *lungs* were not.

In this early skull of the Pangolin the likeness to that of the Crocodile at the same stage is seen in the open pituitary space, and in the cartilaginous nodule attached to each pterygoid bone. But, very soon, in this as in all the other Mammalia, the arrest of the primary cartilaginous jaws to form the incus and malleus (*anvil* and *hammer*), and the new hinge made on the projecting spur of the temporal squama, soon mask the essentially Reptilian character.

In all the embryos and young of the Edentata I find five *vomerine bones*, one along the middle, sheathing the base of the ethmoidal wall, and a front and hinder pair of bony centres.

The front pair, those which are correlated with a pair of retral cartilages growing from the snout to protect the two Jacobson's organs, probably correspond with the pair of vomers seen in the higher Ganoid fishes, in Snakes, and in Lizards.

The hind pair help the middle single bone (answering to the *ploughshare* bone of our skull) to bind together the right and left masses of the ethmoid—the "upper and middle turbinated bones."

In the tooth-bearing forms the hard palate is well developed, and in some Armadillos (*Tatusia*, or the Tatous) the pterygoids help the upper maxillaries and palatines to form this strong floor; a *floor* to the nose, but a *roof* to the mouth.

In the adult Pangolin the pterygoids do not help to form the hard palate; the four palate-bones gape a little at the mid-line, and thus the base of the vomer is exposed. This is seen also in the common Cat, and is quite like what is found also in the Green Turtle amongst Reptiles, and in the Falcons among Birds.

But in the largest and the middle-sized true Anteaters (*Myrmecophaga jubata*, and *M. tamandua*) the hard palate is extended nearly as far back as the base of the cranium, and this is, as in Crocodiles, due to the additional plates formed by the pterygoid bones.

In the smallest kind (*Cycloturus*) these bones are very large, and the posterior nares (hinder opening of the nasal passages) are as far back as in the larger kinds, but the hind part of the floor is formed by a strong fascia of membrane, and not by bone. This small kind gives us the most backwardly placed "basi-pterygoid processes" seen in any skull. The human anatomist will not remember the term "basi-pterygoid process." It was created a few years since for those spurs that grow out from the base of the skull to catch the pterygoid bones in Lizards and many Birds.

In our solid cranio-facial architecture the bones of the upper face have all been drawn up to the floor of the skull, and have become largely confluent therewith, or at least united by immovable sutures. Our internal pterygoid plates form the inner pair of skull flanges, right and left; the outer plates are the external pterygoids, and are mere outgrowths of the larger wings of the sphenoid bone (alisphenoid). In a Sheep these outer plates grow from the sphenoid, much nearer the mid-line, and nearer still in the Guinea-pig; in that type the pterygoid bones, or inner plates, are carried on those inwardly placed outer spurs—as in Lizards and Birds. All these things are specialisations, very exquisite, but not belonging to the deep things of morphology. In the Ant-eater it is the basal part of the occipital ring which gives off these spurs.

The contrast in length between the skull of the great Ant-eater and that of the Sloth is very marked indeed; one is the longest in the Class, and the other is almost the shortest.

"The wisest aunt, telling the saddest tale," is not a more solemn creature than the Sloth; and deep in his cranio-facial organisation there is a peculiar structure, one among many of the strange things in the structure of this joyless Puritan of the primæval forest.

In us, as is well known, the "internal pterygoid plates" develop a retral hook, the "hamular (hooked) process," and the muscle acting upon the soft palate runs its line or tendon over this hook. In one kind of Sloth this pterygoid bone is very large indeed, and is hollowed out into a cavern, into which and from which the breath enters, but has no free escape behind; these caverns run right and left, and each of these may be compared to the "antrum highmorianum" of man. This cavity in the pterygoid of certain Sloths is a structure similar to that which is seen in various parts of the skulls of the higher vertebrata.

As for the use of these caves, in which the wind sleeps for a time in this sleepy creature, I am somewhat doubtful.

One great unlikeness of the Sloth to the other Edentata, especially the toothless kinds, is seen in the deep malar (cheek) bone; in the Armadillo this element is of the medium size, in the Pangolin and Ant-eaters it is either suppressed or greatly aborted. In the toothless kinds the temporal and masseter muscles (those that act upon the parts round the mouth) are small, the grinding action not taking place; thus these muscles do not require to have much force.

I have not only followed the development of the skull in the toothless Pangolin but also in the tooth-bearing Armadillo. In the embryo of the Pangolin the first, or cartilaginous skull (chondrocranium) is feeble, but in the Armadillo it is as well formed as in the Crocodile, or even as in the Frog or Toad; this is soon crusted over with radiating films of superficial bone. The cartilage, as in us, also becomes bony, then the superficial and the deep skeletal elements become largely fused together to form a sort of ivory-box, with a front extension, forming the organ of smell, and with recesses, below, supporting the parts of the mouth and throat. As no tooth-pulps are formed in the Pangolin, the bones of the jaws are not hollowed out, but are very similar to what is seen in the embryo of the Bird, especially of one of the lower kinds—the Ostrich, &c. This flatness of the jaw-bones makes the skull, even at this early stage, very dissimilar to that of the other Mammalia, where the bones swell out with their growing tooth-pulps and teeth.

I have already mentioned that a change, essentially equivalent to metamorphosis, soon takes place in this little skull. This is mainly seen in the inferior arches forming the skeleton of the mouth and throat, and is the last degree of specialisation undergone by a Vertebrate in this part of its organisation. I have already spoken of this change in the non-placental types (those whose embryos are born in a very immature state), but as the process is greatly misunderstood by most anatomists, I will again describe it in this type, one of the best in which to trace it.

The deep and superficial parts that form the jaws in all types, except the Mammalia, are so arranged that the upper part of the first visceral arch forms the upper jaw, whilst the lower segment swings on it as the mandible, or lower jaw. When the first gill-opening, namely, that between the top of the first and the top of the second arch, instead of developing a rudimentary gill, is formed into a pouch opening inwards (a rudimentary *drum cavity*), then the topmost segment of the second, or hyoid, arch fits itself into a small opening in the wall of the ear-capsule on the side of the "vestibule." I am now using terms of human anatomy. The opening formed by the *dehiscence* or splitting of the wall of the capsule is the *fenestra ovalis;* the operculum or lid, which is taken from the top of the tongue-arch or "*hyoid*," is the stapes (or *stirrup*).

This is not stirrup-shaped until we get some height in the mammalian class; in many lower forms, and in all the oviparous Vertebrata, it is a little column, with a broad, flat, oval base, a columella. This may be composed of from one to five segments, but the homology is the same throughout; it is merely the swinging-piece, or top segment of the second, or hyoid, arch of the face, and, as low down among the types as the Newt or the Salamander, is used for a *new* physiological purpose. But the addition of two new segments to the middle ear-chain in all mammals takes place by the arrest and modification of the hinge-piece of the upper jaw, and the corresponding part of the lower jaw. Thus, the amputated lower jaw has to fasten itself to a new swinging point; that, of course, is further forwards than in Birds, Reptiles, and Amphibia, in which this transformation does not take place.

In the smallest Pangolin embryo a little trowel of bone is formed over the temporal region, just as in a Chick; and, as in that bird, the handle of this small trowel projects forwards and downwards over the cheek.

This well-known *zygomatic process* of the temporal bone is a mere spur for muscular attachment in the Fowl; but in the Pangolin it flattens out at the end, and acquires a small patch of cartilage beneath. This is the articular cartilage of the "glenoid cavity" (or socket for the lower jaw); the "ramus" or lower jaw reaches up to this part, covered here with articular cartilage; a *meniscus* of fibro-cartilage intervenes to serve as a "buffer" in the act of eating.

In this embryo most of the rudimentary lower jaw is cartilage, but a thin film of bone is forming on it; yet the coronoid, articular, and angular regions are all cartilaginous. This is the superficial or secondary mandible; but inside it there is a solid rod of cartilage

("Meckel's cartilage"), which is the lower portion of the primary or *deep* mandible or lower jaw of the Reptile or Bird.

The upper part of the primary jaw becomes the malleus (*hammer*), and the swinging-piece, called "quadrate" in the oviparous types, becomes the incus (*anvil*). The topmost segment of the second or hyoid arch is, as I said before, the stapes (*stirrup*).

In the embryo of the Pangolin and Sloth, the stapes is a solid cartilage; in the latter it acquires a small hole after it becomes bony; it is solid also in the Anteaters.

In Armadillos it is *stirrup*-shaped from the first; thus showing a greater nearness to the higher Mammalia, whilst the others show a nearer affinity to the Monotremes, or Prototheria.

I shall describe the peculiar transformation of the jaws more at large when I come to the Insectivora (the Mole and his relatives).

ADDENDUM TO LECTURE IV.

From among the various works on Zoology and Anatomy that appear from time to time I may make a few references, especially from those that relate to the forms treated of in the foregoing Lecture.

I shall merely give the titles of memoirs and papers that have been of most service to me in my own recent researches into the structure and relations of the Edentata.

Professor W. H. Flower, LL.D., F.R.S., Pres. Z.S., read, at the Zoological Society, on April 18, 1882, a very important paper, entitled "On the Mutual Affinities of the Animals composing the Order *Edentata*;" this is to be found in the *Proceedings* of the Society for that year and month, pp. 358–367.

The structure of the skeleton in that Order will be found described in a valuable work by the same author, namely, his *Introduction to the Osteology of the Mammalia* (London: Macmillan & Co., 1876); and his article on the Mammalia in the new (9th) edition of the *Encyclopædia Britannica*, vol. xv. pp. 347–457, will be found of great service to the reader.

The late Prof. Garrod, M.A., F.R.S., gave a paper "On the Anatomy of *Tolypeutes tricinctus* in the *Proc. Zool. Soc.*, Feb. 19, 1878, pp. 223–230, figs. 1–3.

A series of papers by the late Dr J. E. Gray, F.R.S., which appeared from time to time in the *Proceedings of the Zoological Society*, will be found very useful, especially as here and there excellent coloured figures of the various types are given.

I have found the following papers by that laborious zoologist of great use to me in working at this group:—

1. "On the Structure of the Pelvis of *Chlamyphorus truncatus*," *Proc. Zool. Soc.*, Jan. 13, 1857, pp. 8–9, figs. 1–3.

2. "Revision of the Genera and Species of Entomophagous Edentata, founded on the Examination of the Specimens in the British Museum," *Proc. Zool. Soc.*, April 8, 1865, pp. 359–386.

N.B.—In the last two pages of that paper the Prototheria (*Platypus* and *Echidna*) are still classified with the Edentata, although their very much lower nature must have long before been known to Dr Gray.

3. "Notes on the Species of *Bradypodidæ* in the British Museum," *Proc. Zool. Soc.*, May 2, 1871, pp. 428–449, plates 35–37, figs. 1–6.

4. "On the Short-tailed Armadillo (*Muletia septemcincta*)," *Proc. Zool. Soc.*, April 21, 1874, pp. 244–246, plate 41.

Professor Huxley, LL.D., Pres. R.S., in his *Anatomy of Vertebrate Animals*, pp. 330–341, gave the student a succinct and useful account of the structure of these types.

Also, in his paper on "The Evolution and Arrangement of the

Mammalia," *Proc. Zool. Soc.*, Dec. 14, 1880, pp. 649-662, the Edentata are noticed.

Dr Murie, F.L.S., F.G.S., &c., amongst his numerous and valuable contributions to the anatomy of the Mammalia, gives us a good contribution to the anatomy of this order in his memoir on "The Habits, Structure, and Relations of the Three-banded Armadillo (*Tolypeutes conurus*, Is. Geoff)," *Trans. Linn. Soc. Lond.*, vol. xxx. pp. 71-132, plates 20-26.

A description of the shoulder-girdle and sternum of various Edentata is given by the writer in his memoir on those parts of the skeleton in the publications of the Ray Society for 1868, pp. 199-207, plate 21-23.

Dr Philip Lutley Sclater, M.A., F.R.S., Sec. Z.S., who has earned the good-will and gratitude of all zoologists, gives, in his reports on the additions made from time to the Society's menagerie, some very useful contributions to our knowledge of the Edentata.

1. The Cape Ant-bear (*Orycteropus capensis*) *Proc. Zool. Soc.*, June 24, 1869, pp. 431-432 (with woodcut). This is said to have been "purchased on the 18th of June for £150, and believed to be the first specimen of this singular Edentate ever brought to Europe alive. This animal had been purchased at Port Elizabeth, Algoa Bay, and brought to this country by the captain of one of the Union Steamship Company's vessels. It fed well, principally on raw meat pounded small, in the same manner as the American Ant-eaters (*Myrmecophaga jubata*), two specimens of which, obtained in October and November 1867, are still living in good health in the Society's menagerie."

2. In the Secretary's Report, read on the 1st of November 1870, reference is made to the only other species of Cape Ant-eater known at present.

"A male specimen of the Ethiopian Ant-bear (*Orycteropus æthiopicus* of Sundevall). This animal has been placed in the gardens, in company with the Cape Ant-bear (*O. capensis*), received 18th June 1869. The two animals, although both males, live sociably together, and enable a comparison to be made between the external appearances of those two disputed species."

"Duvernoy has already enlarged upon the differences between the skeletons of the two forms (*Ann. Sci. Nat.*, ser. 3, vol. xx. p. 188); and, as will be allowed by everybody, the comparison of the living specimens seems to confirm their distinctness. The chief noticeable differences in the living animals are the more hairy body, especially on the lower back and flanks, the shorter thicker tail, and the shorter head and ears in the *O. capensis*. The insides of the nostrils at their openings are thickly covered with hair in *O. capensis*, which is not nearly so much the case in *O. æthiopicus.*"

3. *Proc. Zool. Soc.*, June 20, 1871, p. 546, plate 43. The sixteenth creature in the Secretary's Report is:—"A Tamandua Anteater (*Tamandua tetradactyla*, Linn.), from the vicinity of Santa Marta, purchased May 29. The clever drawing of Mr Keulemau's, which I exhibit (plate 43), will serve to give an idea of the external form of this animal, which has never been previously received alive by the Society, though we have at present two fine examples of *Myrmecophaga gigantea* living in the menagerie, and have twice received living specimens of *Cycloturus didactylus* (see *Proc. Zool. Soc.*, 1865, p. 385, plate 19). Our Tamandua measures as follows:— Long. corp. 20, caudæ 20, total 40, poll. Ang.

4. *Proc. Zool. Soc.*, June 19, 1877. "A Pangolin (*Manis tricuspis*), purchased May 24 from Mr Cross of Liverpool, being, as far as I know, the first example of this remarkable form of the Edentata that has ever reached us alive.

The animal, which I regret to have to add, died on the 27th ult. from debility, consequent upon ulceration of the tongue, is a male, probably not quite mature. It measures 28½ inches in length, the body being 13½ inches, and the tail 15 inches long; there are seven series of scales on the head, twenty or twenty-one on the back, and thirty-eight on the tail (see woodcut, p. 531).

5. *Proc. Zool. Soc.*, June 14, 1878. The Secretary exhibited a young specimen of Temminck's Manis (*Manis Temmincki*), which had been brought from Zanzibar by Mr Frederick Holmwood, assistant political agent at Zanzibar, and made the subjoined extract from a letter of Mr Holmwood referring to it:—

"The mother of this little Pangolin came from the coast opposite Zanzibar, latitude 6° S., but I have seen what I took to be the same animal, both in Somali-land, under the equator, and as far south as the Makua country, opposite Mozambique. They always appeared to burrow in hard or stony ground, and I saw them always in the daytime. The mother of the specimen I send you lived three months in Zanzibar. She only fed at night, and remained curled up in a ball all day. She regularly retired to the dark corner of my harness-room at daylight, and left for the garden at sunset. There were very few ants; but she seemed to get plenty of insects. She burrowed at intervals all round the garden walls, but this was evidently only to try and escape, as she never made a hole large enough to give cover. The day she had young she came out during the day; but not being quite up to grubbing for insects, she went into the stable and remained among the horses grubbing in the dung. After the birth, she tried to entice the young Pangolin to suck (apparently), sitting up like a dog when begging, and coiling up the moment she got it in her lap. I could not, however, detect whether she managed to suckle it; indeed, I was quite ignorant of the habits of the animal in its natural state. The first day, the young one had soft scales, but they hardened the second day, and it died the same night. The mother wandered about for two days afterwards, and then came into the house and died."

The young one here spoken of was afterwards given to the writer by Dr Sclater; like the nearly ripe embryo of the Aard-vark, kindly put into my hands by Professor Flower since the delivery of this lecture, it was very large for its age, and I suppose that both these Old World Edentates are uniparous. Anyhow, their young when born are relatively stronger than lambs, calves, or foals.

Mr E. W. White, F.Z.S., in the *Proc. Zool. Soc.* for 26th Jan. 1880, pp. 8–11, has given a most interesting and valuable account of the habits of the smallest of the Armadillos—and of the Edentata generally—the *Chlamydophorus* (or *Chlamyphorus*); unfortunately it is too long for insertion here, and too good for mutilation.

Professor Flower, in his valuable article, *op. cit.*, gives a bibliographical list, which includes works on fossil as well as on recent Edentata; and the reader is referred also to Professor Huxley's memoir on the Glyptodon, *Phil. Trans.*, 1865, pp. 31–70, plates 4–9.

Professor Owen's magnificent work on *Mylodon robustus*, London, 1842, may be referred to: his description and figures of the Glyp-

II

todon, and of other Edentates, will be found in the Osteological Catalogue of the Museum of the Royal College of Surgeons, England, 1845.

We owe two of the most valuable and beautifully illustrated memoirs of the gigantic fossil Sloths to the late Professor J. Reinhardt—(1) On Coelodon (Copenhagen, 1878), and (2) On Cryptotherium darwinii (Copenhagen, 1879).

Since the delivery of this lecture, my time, for several months, has been mainly spent at the study of this same group—the Edentata. I confess that this later work has greatly intensified my convictions as to the soundness of the foundations on which we, as Darwinians, are building; if we are patient, the mists in which we have worked, in this the morning of our labours, will, I am satisfied, all vanish, after a time.

Moreover, after a while, wars will cease; now, each man whilst building on the wall has also to be girded with his weapon; with one hand he builds the wall, and with the other he holds his weapon of defence; this is a great hindrance to the work.

The example of our great leader, our most trustworthy Darwin, was perfect whilst he was yet with us. His deeds, if not his words, when tempted to hold parley with his enemies, said—"I am doing a great work, so that I cannot come down."

When this part of the work is done,—when the Edentata are worked out,—and worked into their own side of the great wall, then I feel certain that those who afterwards look at the bulwarks, and tell the towers of the building, will be struck with the strength and the beauty of that part; it will stand out to the eye like a fine flying buttress.

I cannot, here, go into details as to the structure and structural relations of the types that yet remain to us of this Order, or of their relation to the gigantic, recently extinct, forms.

If no fossil remains of Edentata had been found, then any speculation upon the forefathers of such strange creatures might have seemed unscientific and useless, showing rather the self-confidence of the speculator than the value of his thinkings and imaginations.

Here, however, Nature herself leads us on, and bids us not fear nor be faint-hearted. Time, in his hurry to garner the last great fruitful harvest of these types, only partially hid some of the most precious of his spoils.

Things at their height are ready to decline. The Edentata undoubtedly arose to their height or culmination, as a group, at the end of the ages just before our own time; but their decline has been fearfully rapid. Judging from the size of the pelvis of the Megatherium, and, remembering what has just been said about the large size of the new-born Aard-vark and Pangolin, we may suppose that the young of that giant, at birth, must have been very large indeed. So large, I suppose, that if, indeed, the *old* and the *new* Sloths co-existed, the young of the former would seem fit to be the parents of the latter; they—the *Ai* and the *Unau*—although adult, would have looked mere kittens beside the huge suckling of the Megatherium.

There is, however, much more to be spoken of in this matter than the mere size of these extinct types as compared with their dwarfed modern representatives.

The Megatherium and the Mylodon were not *pure* Sloths; the Glyptodon was not a *pure* Armadillo; they were more generalised than the lower forms, and to understand them, we must compare them, not with the living kinds in their adult state, but in their embryonic stages. This is the special, and not easy, work of the morphologist; a piece of work, on these very forms, is likely to see the light in the coming winter, and the reader may then take up the details which will be laid before him. The features of the Magatherium may be seen in miniature in the face of an embryo Unau; but, more than this, that embryo shows, most unmistakably, a likeness to the family of the Ant-bear. The latter, the longest-faced creature living,—that gigantic mammalian Weevil, with his hairy body, and curiously out-drawn long slender face,—has a shorter-faced relation not larger than a Squirrel: this is the climbing Little Ant-eater, with a prehensile tail like that of a Spider Monkey. That kind, whilst still young, has a skull which has most remarkable points of resemblance and *affinity* with that of the early embryo of the Two-toed Sloth (Unau, *Cholœpus*), which, *then*, has a longish face, whereas when adult, it is curiously and almost absurdly short. So that I can read some tokens of the Ant-eater, as well as of the Megatherium, in the nascent face of the existing Sloth. Mark another point well

worthy of mention. It is not until we are some good height in the
mammalian scale, that we find the mechanism of the ear perfected by
what is called a *stapes*, or stirrup-bone, in the middle-ear, fitting on
to the *vestibule* of the inner or essential ear. The oviparous tribes
which have that peculiar operculum or covering to the oval opening
of the vestibule (*fenestra ovalis*), namely, Amphibia, Reptiles, and
Birds, nearly always have it stalked, so that it forms a little column
(columella), with a dilated upper or inner end. The Prototheria, and
several of the Metatheria, have this columella instead of the stapes or
stirrup-shaped element in the middle-ear; in the Edentata the
Armadillos and the Aard-vark have a stapes; the Sloths, whilst in
their embryonic state, the Ant-eaters, and the Pangolins, have a
columella. Thus, in this transitional condition between the oviparous
and nobler mammalian tribes, the Edentata and the Marsupials are
about on an equality. And this is true all round; for in some things
the latter have the pre-eminence, whilst in others the Edentata are
manifestly superior to the Opossums, Phalangers, and Kangaroos.
If the New World and the Old World Edentata ever had common
parents, a considerable amount of time must have elapsed since, to
give them the chance of becoming so very distinct as we now find
them. If the long-tongued ant-eating Woodpecker is a kind of side
branch from the primary Passerine stock, and the long-tongued
insectivorous Chameleon is a sort of side-branch of the Lizard stock,
then we may expect curious things to take place in a mammal also,
which loses all his teeth through taking to live on Ants. I
am arguing as my esteemed colleague Professor Flower argues,
and we are of one mind on this matter. I am also in agreement
with him when I incline to put the Aard-vark a good way off from
the rest of the Order; it comes nearer the Insectivora than any
other member of the group, whilst the Pangolin refuses to acknow-
ledge more than general relationship with the forms from the New
Tropics.

Yet the Pangolin has an equal right with them to be considered a
descendant of some prototherian beast—some common ancestor to
him, to the Neotropical forms, to the Duck-billed Platypus, and to
the ant-eating spiny Echidna.

All the better sort of Mammalia have a plate of bone dove-tailed
in between the great skull-bones (parietals) and the back wall of the

skull (occiput). But none of the Edentata have this piece except the Aard-vark, although the Marsupials agree with higher forms in having this superadded plate. Now, as far as I can see at present in my dissections of the young of the Platypus and Echidna, the parietals, as in Lizards and Snakes, fuse together very early, and do not keep apart, either for a long time, or for the whole of life, as in noble mammals; their back-skull is very large, and turns over the brain-cavity; it is roof as well as wall. In the Edentata, generally, this early reptilian fusion of the parietals does not take place, but they have a huge back-skull, which helps the parietals, without an intercalary interparietal to finish the skull, above. The Pangolins, whose arrested covering of hair degenerates, so to speak, into a quasi-reptilian condition, have, in some species, a breast-bone with long, hinder, cartilaginous horns, like *Stellio* among the Lizards. Also, on the right side, in the abdomen of this unsymmetrical creature, there are four cartilaginous abdominal ribs, like those found in certain Lizards, namely, *Chamæleo*, *Polychrus*, and the archaic New Zealand *Hatteria*.

More than this, in the mammals whose embryology I have studied, I have never found such evident marks of degradation of the primary or cartilaginous skull as in the Pangolin; the wonderfully specialised and peculiar skulls of Serpents, Lizards, and the Tortoise tribe, are the only other instances in which I am familiar with the stoppage of growth of such a primary and important structure as the inner wall of the brain-case. This remarkable fact, whilst it suggests some degree of degenerative change, in no wise leads us to suppose that the Pangolin, the Snake, and the Lizard, are in any way nearly related, now. They, each and all, after separation from the main old root-stock, got into their own grooves; they improved in some things, and got a little way backwards in others; they have not continued as they were since the creation of the world, but have suffered from the mutability of all things on this planet; and at times, like Man, in his higher sphere, when they were not improving, they were degenerating.

In all the endless modifications of animal forms we see that the morphological force has ever been looking towards two ends in each individual creature, namely, *food* and *safety*.

In the case of our own species, that which is to be desired is, first that a man may eat of the labour of his hands, and then that he may lie down in safety. "Each man under his vine and under his fig tree, none making him afraid," is the favourite image of a simple, but delightful, form of human life. Then each one of these peaceful men may bring home to him, in his quiet nook of pleasant retirement, one who will gradually fill his home with the new and joyous life of a fresh generation.

But hunger and fear have to be got rid of first. "I must eat my dinner," says the semi-human Caliban. "We must sleep o' nights," is the remark of the guilty Macbeth.

Everyone who has watched the habits of the solitary Blackbird or the gregarious Starling will have noticed that they are always on the watch, and that every movement of their most elegant bodies is dictated either by want or fear. That is, before their life has been crowned with the joy of offspring; then, indeed, they for a time cast all their care and caution to the wind, and pour out their souls in gladness. So that, to food, and safety, we have to add love of, and delight in, offspring.

Now the question arises—Have these marvellously active and suspicious birds always had this perfection of bodily structure, a structure in some cases so perfectly adapted to flight that they can almost do the feat of "Ariel"—"put a girdle round the earth in forty minutes?" Or is this the ultimate result of a marvellous correspondence of the organising power with the surroundings of the creature?

The creation of such a creature as a high or culminating type of singing-bird, is worthy of a Divine interposition; but that does not settle the matter—does not answer the question.

Of this I am certain, that Nature has not failed of a grand purpose if she has succeeded in bringing such a type to perfection during the ages that have elapsed since she flooded the old representatives of the modern cane-brakes, that they might reappear, in our time, as coal. That, certainly, is a long period for the modification of an organism, but Nature (or Wisdom), always rejoiced in the habitable parts of the earth, and Her delights, prospectively, were with the sons of men. The analogy of Nature, here, is in perfect harmony with the poetry of the East; her noblest daughters love their offspring long before their eyes are gladdened with the actual sight of them.

But Nature has not worried or fretted herself over her work. Her progress has been as calm as it is purposeful. Of this we have a homely illustration in the inimitable *Water Babies* of Charles Kingsley. With fine instinct, Kingsley caught the genuine spirit of modern Biology, and rightly judged that children should be indoctrinated with it. One short quotation will serve our purpose here. Tom, the Water Baby, comes to Mother Carey's shrine (Mother Carey, we need hardly remind the reader, is the name—not a very dignified one, it must be owned—applied by our poet to Dame Nature). The little man approaches with awe and wonderment, expecting, of course (like some grown people who ought to know better), to find Dame Nature "snipping, piecing, fitting, cobbling, basting, filing, planing, hammering, turning, polishing, moulding, measuring, chiselling, clipping, and so forth, as men do when they go to work to make anything. But, instead of that, he finds her sitting quite still, with her chin upon her hand, looking down into the sea with two great, grand blue eyes, as blue as the sea itself. Her hair was as white as snow—for she was very old—in fact, as old as anything which you are likely to come across, except the difference between right and wrong."

"I heard, ma'am," says Tom, "that you were always making new beasts out of old."

"So people fancy. But I am not going to trouble myself to make things, my little dear. I sit here and *make them make themselves*."

To return from this digression. Between the morphological force within, and the forces of nature in the external surroundings, there is, if one may so speak, a balance struck. Thus, everywhere in nature, all things are double; one thing is set over against another; and forces, apparently antithetical and antagonistic, in and by their very struggle, produce the most exquisite and perfect results.

In the Edentata no perfection of special modification redeems them from mammalian lowliness; they are the slow, dull, heavy-gaited churls of the class to which they belong, whilst the sharp-eyed cat, and her sharper owner, are two of the highest and most perfect forms in the class. What is it that lies at the root of this difference? I answer, the relative development of the central nervous system—the organ of the mind. A similar difference to that which we note between the Edentata on the one hand, and the boy and his cat on the other, also exists between these two types of the higher Mammals, of which the singing bird is always in fear—namely, that whilst the cat has in many respects a much more specialised bodily structure than the boy, yet he far outdoes her in the capacity, and energy, and marvellous attributes of his central nervous system.

Now, returning once more to the Edentata, what do we find? We find that everything has been done for their safety that could have been done, and yet they have not been safe; they are fast becoming extinct, and the greatest and noblest of the Order have already been elbowed off the planet. But in the Cat and her relatives, and, above all, in the human race, every danger has been despised; they have been specialised for conquest, not for passive resistance and cowardly meekness. So that it is very probable, that when even Man himself becomes mastered by his surroundings—when Campbell's "last Man" stands and addresses the dying sun, his faithful Cat will be erecting her tail, and softly purring at her master's legs, whilst he utters this his last speech. Now, during all the changes and chances of time in which the Edentata have held their ground and kept their place, what is it that has wrought them into such strange, and to us grotesque, shapes—shapes, however, that are very admirable indeed, when the life and habits of these creatures are considered? As students of nature the less we mystify ourselves with metaphysical speculation the better; the theologian very properly bids us not to "tread on holy ground." We have, therefore, only to deal with what are called second causes; we are only competent, as biologists, to deal with these. Here, however, on our own ground, we are not tethered or limited.

Lord Bacon still speaks to encourage us in our research, for he smartly says—"It is good to ask the question which Job asked of his friends:—'*Will you lie for God, as one Man will do for another, to gratify him?*' For certain it is that God worketh nothing in nature but by second causes; and if they would have it otherwise believed, it is a mere imposture, as it were in favour towards God; and nothing else but to offer to the author of Truth the unclean sacrifice of a lie."

If it is difficult for any one unused to biological research to imagine how such diverse forms as the Ant-bear, the Sloth, and the Armadillo sprang from one common *root-stock*, and became so diverse in form and habits in answer to the special needs and conditions of each kind,

I would refer him to an instance much more easily grasped and understood.

Southey's little poem, beginning with—

> "Oh Reader! hast thou stood to see
> The holly-tree?"

is an exquisite piece of still-life Darwinism, notwithstanding that the writer of it, had he lived a generation later, would, I believe, have been the last to become a convert to Charles Darwin's theory. Therefore that contribution to the doctrine of development is all the more valuable, as being utterly undesigned.

Yet the *polymorphism* of the leaves of that beautiful tree admits of but one interpretation; they each and all have responded to their surroundings, setting up their prickly backs, like Hedgehogs, in the lower parts of the tree, lest the ox that licketh up the grass should lick them up also; but above, in the steeple-like culmination of the tree, right under the eye of heaven, the defensive prickles are suppressed, and each leaf glistens in the sunshine, unarmed and void of fear. Was each leaf separately created in spring-time? You answer—"No, no need for that; the forces within the tree, working in exquisite harmony with the surroundings, sufficed to make all that difference in the form of the individual leaves." I rejoin—"Are you assured of that? if so, good! We now can understand each other." Of course every observer of nature is acquainted with a thousand instances of the same kind as that presented to us in the holly-tree; yet these familiar phenomena all speak one language;—that language is no longer barbarous; everyone understands it now.

That lies outside, but illustrates, our work; every member of the animal tribes merely lives out the cycle of an individual life, which life is one continual struggle against drought and rain, heat and cold. Of course, each kind has the benefit of the whole accumulation of excellences developed in its own direct ancestry; each oak tree and holly tree enjoys the rich inheritance, so also does the Armadillo, the Ant-bear, and the Man. I have no doubt of one thing, namely, that Armadillos, like Tortoises, are the descendants of types that were not cased in complete armour. I have, likewise, no doubt that Ant-eaters and Pangolins, like Tortoises and Birds, are the descendants of types that had a perfect series of teeth. The Ant-bear's head and tongue have lengthened through the ages; in this he has gone on unto perfection. The Sloth, also, has cut off, not his right hand, but all his unnecessary fingers, during the long secular period in which he has been slowly moving towards the mark of his particular type of excellence, and his face has gone on shortening, so that, although a long way below a Monkey, there is something monkey-like in his curious short muzzle.

Time, that unwearied harvest-man, has almost finished his mowings in this field. Here and there, his two-handed engine has left a patch of poorer stuff, a narrow headland of weaker, and later growth. The rest he has garnered and locked up safely, not, however, without letting fall here and there, as if by accident, a handful or two, which the stranger from a far country has been too glad to glean. But if Time has been so much against the modern biologist, there was no reason why Nature (or Wisdom, to use another image) should have been so anxious to hide these types from us; nothing more perfect was ever developed by the morphological force. Here one can speak great words, and yet not be hyperbolical; the little finger of the Megatherium was literally bigger than the sloth's loins, and all the existing Armadillos (one of each) might be packed in the body-case of a Glyptodon—the extinct Armadillo with sculptured teeth.

The largest Green Turtle would look a poor, shrunken thing beside a living Glyptodon, a mean hero in a mean armour, not worth ten oxen, but the Glyptodon's armour could not be bought for a hundred oxen. Yet that reptile-in-armour, in his own element, is like a swift ship; the Glyptodon had to carry his superb body-house on stumpy legs; he must, in walking, have seemed like a huge bombard full of liquor, that was being carried off by his oaken tressles or undersetters.

If the Megatherium, or his somewhat more modest-sized relation the *Mylodon* (another extinct Sloth), did find their supply of food in the way palæontologists suggest, their mode of dining must have been a sight worth seeing. That delightful, typical Englishman, the late Rev. Sidney Smith, once reviewed Waterton's *Wanderings*, and described the strange, grotesque, weird, unthought-of creatures of the New Tropics. Would that he held the pen here, now!

Let us, however, try to imagine a Megatherium waking up after lazily dozing a month or two during the dry season, and then, hungry and wet, in the heavy downpour of the beginning rainy season, setting to work to break his fast. As far as can be judged by the tools he had to work with—paws a yard, and claws a foot, in length—the first thing to be done was to throw out a few hundredweights of earth from the roots of some large tree.

Now he changes his tactics; he has good collar-bones, and well-shaped arms for embracing; so, bear-like, he hugs the tree upon which his desires are set, and, busily digging still, not now with his fore, but with his hind, paws, his great weight resting upon his haunches and his tail, he, with many groans, sways the big tree to and fro; at last with a great crash it falls, not, however, without giving him some sense of its weight, for it was a tree worthy to grow in a forest trampled upon by this atlantean Sloth.

That large crack in the outer table of his skull is of no consequence; his small brain is a long way off, and there are many empty cavities to be found in a head like his; those broken tiles over the empty spaces of his head will soon be mended, and what would be *pain* to us, is to him a pleasant sense of tickling.

But Sloths live in trees, climbing from branch to branch, supine, with strongly bent wrists and hooked fingers! Yes, I admit, such Sloths as live in these degenerate days, but not the Sloth I am speaking of. Think, if you can, of a Sloth, half as thick again round the waist as an Elephant; with a tail as bulky as a dray-horse's chest; and feet as large as the many-knotted roots of the gum-tree; think I say, of such a Sloth climbing trees. No, it is your poor little dwarfed modern Sloth who climbs—not the large Megatherium.

Our gigantic *Prospero* has plucked up his cedar by its spurs; his millstone-like teeth—he also is a *Mylodon*—and strong jaws will do the rest; he need not hurry and he will not. He has "blessed his maw" to this good hour, and will now enjoy himself. The sorrow of it is that he is not to this day digging up, pulling down, and eating, the trees of the forest, for us to see the sight. For death has gnawed upon these huge beasts; they are laid in the grave—their eternal dwelling.

Speaking of such an one going down to the nether parts of the earth, we are reminded, by way of contrast, of the poet's account of the slow but sure *fossilisation* of the rude forefathers of the hamlet—

"Each in his narrow cell for ever laid."

Seven feet of earth is enough for the biggest of these undeveloped, mute, inglorious Miltons of the little lone country place.

When, however, the General Sexton gave the Megatherium his *seven* feet of earth, this was merely the measurement crosswise, for he was nearly that breadth across his loins with his flesh and his fell on; his length was double that and more. If time had not failed me, I would have described his parts and his power, and his comely proportion. I might have dilated upon his feet, to the extent of filling a chapter; the spoor which they made, being filled by timely rains, made a pond in which the Axolotl might have disported, and in which the largest of Frogs did, undoubtedly, take their pastime.

If the reader will visit the great Natural History Museum at South Kensington, and the Hunterian Museum in Lincoln's Inn Fields, and look at the remains of the fossil Sloths, he will make this discovery, namely, that the writer's attempt at the rejuvenescence of these beasts is dull and flat, and that not half the truth has been told him.

LECTURE V.

INSECTIVORA.

THE Edentata are very similar to the Insectivora, but are arrested and modified largely on their own level. They are more related to the Monotremes than to the Marsupials, but the Insectivora appear to be rather a development or outcome of the Marsupials. The works or papers that have been of most value to me in dealing with the Insectivora are by Professors Huxley and Flower, and Dr Dobson; we are rich, however, in the literature of this Order. It has taken me many years to collect materials for my researches into the development of these animals. I published some of my results long ago—those which related to the shoulder-girdle and sternum—but the accumulated matter on the skull, work done during the last two years, has not yet seen the light. In working out the details of the development of the skull, I have followed my own bent, gladly accepting help from my fellow-workers in biology.

But in arguing upon what has been seen and registered in this special piece of research, I have been largely influenced by one whose aptitude for drawing deductions from facts, and for putting those deductions into such a form that other minds can receive and appreciate them, is far beyond anything I can boast of. The author I refer to is Professor Huxley, and the paper entitled "An Application of the Laws of Evolution to the Arrangement of the Vertebrata, and more particularly of the Mammalia," has been repeatedly referred to in these Lectures; I have had much help from him, and from other fellow-workers. But I cannot pass unnoticed Dr Dobson's valuable work on the *Anatomy of the Insectivora*, nor Professor Flower's important papers and works on the Mammalia generally. Although I shall refer merely to what can be seen in the skull and face, I have not been unmindful of the rest of the organisation of these creatures; but my views are mainly based on what can be seen in the head. Now the types treated of in my last lecture—the Edentata—as I showed, lead nowhere; they end in themselves; not so the Insectivora. In them we evidently have the modified and dwarfed representatives of the original placental mammals, or Eutheria. These lowly forms, although small and inconspicuous, yet yield a rich harvest to the biologist, for they are the somewhat altered, living patterns, of the forms that did abound in the middle, and even in the early, Tertiary epoch. In the Secondary rocks their existence is doubtful, as the bony remains of the earliest Insectivora would be scarcely distinguishable from those of the Metatheria or Marsupials. Once beyond these lower forms, however, we soon find ourselves in the midst of types, which, if very unlike our modern Insectivora, are yet much more unlike the higher forms of beasts now existing. The four- or five-toed feet, the simple tooth-pattern, the generalised condition, indeed, of all the parts, and the very small brain cavity seen in their remains, show us that we are only just above the Metatheria. The commonly received opinion of the multitude is, that the first-recorded beast-namer was contemporary with all these extinct forms, and also that they were his, and that he put either his brand, or his ear-mark, upon them all.

All the evidence lies the other way. There is every reason to believe that that first zoologist was familiar with the beasts whose forms are so well known to us now—Lions, Bears, Horses, Cows, and Sheep. But such large and small cattle as he was, and as we are now, familiar with, were not to be seen in the days of the years of which we speak. In those days, no shepherds kept watch over their flocks by night—Sheep were not, and the paw of the Lion and the paw of the Bear had not been developed. It is evident that we must, in imagination, wait for a great time whilst the earth is preparing for these culminating types; for some twelve thousand feet of fossiliferous and other rocks have gradually been laid down since the first Eocene types of mammals appeared. The mammalian forms of that time are more nearly related to our existing Insectivora than to any of the groups that lie on a higher platform or level. From some such common root-stock one great sucker or stolon after another arose:—Rodents, Bats, Hyracoids, Proboscideans, Herbivora, Sirenia, Cetacea, Carnivora, and Primates. At first sight it might seem that the huge Whales were more worthy to have been brought upon the scene miraculously than the little Bats, but as both these sorts of beasts are continually developing, even now, from an almost infinitesimal pellet of protoplasm, it seems to me that the large types came into being as easily as the small. Speaking of protoplasm, I may remark that the mind is overwhelmed when it attempts to follow the embryology of one of the larger Cetacea; however its feats are accomplished, it must be confessed that protoplasm has an amazing power of growth. Whatever may be the difficulty as to the gradual modification of a terrestrial form into one of these swimming islands, there can be no gainsay to the fact that each living Whale repeats its ancestral history in its own lifetime, more or less. Anyhow, in this present period, Whales exist, however we may account for their existence. Can the lowly Insectivora throw any light upon the evolution of these huge types? In attempting to answer my own question I shall speak mainly of those diagnostic characters which are to be found in the head; these will be looked at both from below and from above, as they show ancient or modern relationship. The teeth, of course, are amongst the best of all characters for the use of the taxonomist; if they do not dominate the creature, they are in harmony with all the rest of its organisation.

But these superficial, and easily studied, parts are not the whole of the matter; I should be relieved if they were, as my task would then be much easier. If morphology seems obscure, it should be remembered that it has to do with the obscure corners of nature. Now the Insectivora are so related to the various orders above them that they are for ever anticipating their diagnostic characters, and they are so related to the types below them that they are constantly seen to retain the marks of those lower forms. They are an exceedingly variable group—promiscuous, so to speak—although in their external adaptive characters they are more uniform than might be supposed. I speak now both of the teeth and the limbs; the teeth are of a simple type as compared with what is seen in the higher Eutheria, and the limbs are generally typical as to the number of the digits, which are seldom less than five. Also in the deep, or inner, part of the fore-limbs —the shoulder-girdle—the Insectivora are all typical except one, namely, *Potamogale;* having well-developed clavicles; thus they are capable of using the fore limbs for very various purposes; the pelvis (hip-girdle) is not unfrequently open below.

I

As to apparent uniformity, they are in contrast with the Marsupials, whose outward form and adaptive modification of teeth and limbs are much greater. But in that which lies deeper than teeth or limbs there is evidence, in the Insectivora, that they are a group whose organisation is full and fertile of the power of adaptive change.

Now if we compare our present living Insectivora with the extinct Eutheria of the early Tertiary period, these two Faunæ are manifestly similar; they would indeed form one very uniform group if we could get back again all those types that nature has wasted and buried. If all those hidden treasures of the secret places of the earth, and all those that failed to leave their traces even there, could be restored to us, even, as it were, in "the valley of vision," then we should see that our living Insectivora are only the waifs and strays of countless groups of Pro-eutheria, of many a size and shape, but with very simple tooth-crowns; with mostly pentadactyle feet; with small brains; and with a low intelligence. But as in the rude forefathers of the hamlet we have the quiet and unambitious progenitors of the men who, when their time comes, turn the world upside down, so in those Eocene and early Miocene quadrupeds—the equivalents of our little living Insectivora—we have the rough unhewn forms from which our noblest types have arisen. In those days the Mammalia, generally, had not only a very simple form of tooth-pattern, much like that still seen in the Marsupials, but the teeth were in great number, and often with no interspaces. The digits gradually began to abort; the innermost (the *hallux* and the *pollex*) going first; but no sign was shown then of such feet as we see now in the even- and odd-toed types of our noble, existing, hoofed forms. The Cow, that parts the hoof, cleaving the foot into two equal portions, and the Horse, who brings his springy weight down upon a single digit on each foot—a digit that has drawn the life out of the others—such forms as these had no existence until near to the time when the ruler of the beasts appeared.

In the olden time the term *proboscidean* would have been applicable, not to an Order, but to certain Genera. Even now this elongation of the double nose-tube, and its segmentation into a ringed structure, is not confined to the Elephants, as I shall soon show.

If there were any supra-mundane biologists watching the evolution of forms on our planet at this time they would see the first promise of defensive horns, and the gradual specialisation of certain teeth for offensive and defensive purposes. If we suppose that such watchers of creation existed, and that they had joy at the sight of those strange beasts of an early age, then what must have been their feelings when they saw, at last, the forms with which we are all familiar —Antelopes, Oxen, Deer, Bears, Wolves and Lions?

Nature has, undoubtedly, touched up the form of even the small conservative remnants of that old Fauna; the Hedgehog, the Mole, and the Shrew; the Colugo, the Tenrec, and the Tupaia; each of these has its own style of beauty, and its own most perfect adaptation to its surroundings. That the old quasi-insectivorous types were the root-stock out of which the higher Eutheria arose is made probable by a remarkable fact, namely,—to quote Professor Huxley,—that "numerous Lemurs, with marked ungulate characters, are being discovered in the older Tertiaries of the United States, and elsewhere."

Further, to continue my quotation—"No one can study the more ancient mammals with which we are already acquainted, without being constantly struck with the insectivorous characters which they present. In fact, there is nothing in the dentition of either Primates, Carnivora, or Ungulates, which is not foreshadowed in the Insectivora; and I am not aware that there is any means of deciding whether a given fossil skeleton, with skull, teeth, and limbs almost complete, ought to be ranged with the Lemurs, the Insectivora, the Carnivora, or the Ungulates" (*Proc. Zool. Soc.*, 1880, p. 651).

In severe scientific research it is dangerous to take things upon trust, yet nothing could have been more opportune, to me, than the appearance of the paper I have now quoted, just as I was beginning to work at the development of the Insectivora, and the forms and types immediately below them. I confess that but for the enthusiasm with which that paper has inspired me, I should have been afraid to draw such bold conclusions as the author of that paper draws; yet, in meditating upon the facts that are daily opening up to me in my own especial line of research, the truth of these deductions becomes more and more evident. Now, if these things are true, what is to be done with the old Systems? Where is Linnæus now? and where Cuvier? Where are your old hard and fast landmarks—your stony dykes that kept the types apart?

If there is any one whose happiness depends upon the safe preservation of these old things, to him I have nought to say; for myself, that which is found to be false I should gladly see cast aside and forgotten. No Zoological System was revealed to the first man who named the cattle, yet, I repeat, his cattle were similar to ours, and were not the same as those of the early Tertiary period; in his time there were Swine, Ruminants, and Single-hoofed types. Palæontology has preceded embryology in this field of biological research; in embryology the harvest is great, but the labourers are few.

The existing Insectivora lay their hands, so to speak, both on the low and on the high; they are indeed the connecting links between the higher forms on the one hand, and the low marsupial, and low monotrematous, types on the other. It is hard to say what their own diagnostic marks are, for they are evidently not hardened into a fixed zoological group, but are, as it were, a collection of plastic types; low, as having the stigma of ancientness upon them, and yet full of the promise of all that is highest in the great mammalian class.

In all this group, only one type, the large aquatic otter-like *Potamogale* of West Africa, is devoid of clavicles; also the presence of five fingers and five toes is very constant; the pollex is deficient in *Rhynchocyon* and one species of *Oryzorictes*, and the hallux in *Macroscelides tetradactylus*. *Five* is evidently a sacred number to nature; the Amphibians—Salamanders and Frogs—usher in this fixed number, fixed as against a greater number; and man rejoices in its retention in his own hands and feet. That which often characterises a declining dynasty is the dwarfed condition of its members, as well as the loss of certain of its families; in the past history of the types it is generally written down that "there were giants in those days." In minuteness, one of our native Shrews is the rival of that smallest Bat, the Pipistrelle, and that smallest of the Rodents, our beautiful squirrel-like Harvest Mouse. These three types of Mammalia show us how small a vessel of life will hold all that is essential to one of our own class; these nursing mothers are no larger than some of the insect tribes. Here we see that, as the lowly herbs of the field escape the violence of the storm, when the forest-king is hurled from his throne, so these exquisite little creatures inherit the earth because of their humility and meekness. It is a fact that they, and forms like them, have kept their ground during the ages that have witnessed the extinction of numbers without number of the strong burly-boned giants of the Class. Yet the small Shrew has numerous enemies; the Cat mistakes it for a Mouse—a mistake common enough amongst us—but does not eat it; that feathered cat, the Mousing Owl, swallows large numbers of both the land and water Shrew; she has been my *Falcon*, for through her I have obtained my best specimens. Both the Shrew and the Hedgehog are considered uncanny by the country people; have they an instinctive sense that these are ancient, and even degraded, types? To the biologist there is no form in the group of greater interest than the common Hedgehog, which, specialised highly enough as to its outer skin, is found to be very generalised when studied in its development. Taking the skeleton of the Hedgehog, merely, it is a good example of what is general, rather than special. Free from all violent modifications, it is very useful for comparison both with the old forms of the early Tertiaries, and the new forms of the present period. I have worked much at its skull. It serves as a kind of epitome for the rest; when once you have mastered this you easily see the meaning of any other kind of eutherian skull; by it you can in some degree measure the various kinds and degrees of specialisation to be seen in the other and higher kinds. In the skull of the embryo Hedgehog we see several characters that are familiar to us in that of an embryo Reptile or Bird ; and some are like what are seen also in the Marsupials. As we ascend in the scale of the Orders these quasi-reptilian and marsupial (metatherian) characters die out, more or less.

Beginning with the superficial bones that cover the inner or cartilaginous cranium—besides the flatness of their form, covering as they do, a flat skull with a small internal cavity—we see that the single plate, called "interparietal," is very large. This addition to the occipital arch is peculiar to certain groups.

In my former lectures I have frequently spoken of the gradual specialisation of dermal scutes (or plates of the superficial armour) as we ascend in the scale of the Vertebrata; the internal skull—with its contained organs—dominating the outer parts, making them answer to what is wanted for protection, both in number, in weight, and in measure. Now, in fish and in reptilian forms, besides the main frontal bones, we frequently find a series over the brows, the supra-orbital scales; these linger even in the bird class. As yet, in no mammal except this, have I found more than one frontal bone, right and left; but, for a while, in the Hedgehog, the orbital rim and plate are separate, as a single, distinct, lateral piece. This might be considered to be a very small thing of itself, but many grains make a heap, and facts of this sort, as to exceptional characters, now accumulating, are becoming very numerous indeed. Where the maxillaries and palatines meet in the hard palate, there these bones are deficient to some extent, as in the Marsupials. Above these, toward the mid-line, the vomer and its companion bones are remarkably well developed in relation to the large Jacobson's organs, and so are the retral tracts of the alæ nasi, or cartilages of the snout, that encapsule those remarkable organs.

As in the Bird and Reptile, the pituitary space is open; that is to say, the seat of the turkish saddle has a round hole through it. As a rule, the Mammalia agree with the cartilaginous Fishes and Frogs in having this part filled in with cartilage. The air-galleries into which the drum-cavities open in the Crocodile and the Bird are represented here by a large lateral recess, right and left, the basi-sphenoid giving off a large, hollow wing, which takes the place of the distinct *bulla* of the Cat. Among the various modifications found in the Insectivora this is one of the most constant, but it is not universal. In the embryo Hedgehog, before any superficial bones are drawn towards the cartilaginous skull as its support (a state of things like that which is permanent in cartilaginous fishes), the deep cartilages forming the lower jaws are very solid for a mammal, and have superficial cartilages strapped to them. These superficial cartilages are largest in the Chimæroid fishes; and the deep cartilages lessen in bulk as we ascend into the various culminations of the fishy, reptilian, and avian types. Here, in the Hedgehog, we are much nearer the cartilaginous fishes in this respect than when amongst Reptiles and Birds; in this, as in some other things, even the lower kinds of the Eutheria, or placental Mammalia, show a relationship to the lowest sorts of fishes known. But there are no fishes now living, low enough, or generalised enough, to be at all good living representatives of the stock out of which the highest and best sort of living Vertebrata must have arisen. We must, however, try and be content with such things as we have, and by comparing the early stages of the mammal with even the permanent condition of some of the cartilaginous fishes, we do get some light upon this dark and difficult subject. If you will put together what is familar to you in the parts of the human face, and recollect some of the many things I have from year to year reiterated about the face of Sharks and Skates, you will have a very good idea of the marvellous transformation the mammalian embryo undergoes during development. Meckel's cartilages (the deep or inner lower jaws) are immense in the embryo of the Hedgehog; they meet and unite in front, at the chin, and there form a single bar or basimandibular rod. Just behind this single, terminal, part these cartilages become flattened and very solid, much more like their counterparts in the Shark than what is seen in the more specialised forms of fishes. I have followed the changes that take place in these parts, through about nine stages, for a lesser number would not have given me all I wanted in searching after the meaning of this transformation. Before this rod becomes ossified, a thin superficial plate of bone, attached to, and grafting itself upon, a thick superficial slab of cartilage, appears above and outside the lower two-thirds of Meckel's cartilage. The bone is the well-known dentary of ganoid and bony fishes; the slab of outer cartilage answers to the small lower labial of a common Shark, and to the huge massive lower labial of a Chimæra. I have repeatedly shown that the upper and lower jaw of those kinds of fishes is formed by the bending of the first internal gill-arch over the cavity of the mouth. The upper jaw, then, of a Shark, is, in technical language an "epi-branchial" element, the lower jaw is the "cerato-branchial" of the same first post-oral arch. Above the Sharks and Skates, the joints or segments of the gill-arches become ossified, and each piece is further segmented into two, so that above the epi-branchial we have a "pharyngo-branchial," and below the cerato-branchial there is generally a "hypobranchial." These further subdivisions we may forget for the present; they are very inconstant in the first and second arches of the throat.

Very soon, whilst the superficial bone is forming by transformation of the superficial cartilage, a thick, solid bar of bone is formed in the front third of the Meckelian rod of the embryo Hedgehog. The upper part of the cerato-branchial bar becomes detached after a time, but not until it has become ossified; this ossification is arrested, to form the malleus (or hammer bone) of the middle ear. The huge epibranchial, or upper jaw of the Shark, is represented in the Hedgehog by two tracts of cartilage, one small and the other large.

The large cartilage is the hinge part—the hinder region of the upper jaw. The fore part, which in the Shark carries the upper teeth, is represented by an oval segment of solid hyaline cartilage, which becomes converted into the hamular process of the pterygoid bone; the larger hind piece becomes the incus, or anvil. The stapes is stirrup-shaped, and is, as I have before stated, the pharyngo-branchial element of the perfect hyoidean (or second arch). There is a ring, partly cartilaginous and partly bony, formed round the interspace (cleft, or tympanic cavity) of these two arches. The inner part is bony, and forms the annulus, or osseous ring for the ear-drum; the next is a partly segmented series of cartilaginous annuli or rings forming the coating of the meatus externus or ear-porch, which ends in the concha or projecting part of the ear.

The whole of this latter structure is a specialisation of the familiar "spiracular ray" of the Shark—that small cartilage which supports the membranous flap (operculum) of the blow-hole between the eye and the ear.

There is nothing new in the main part of the skull containing the brain; in the skeleton of the organs of special sense; or in the superficial parts of the head—snout, lips, outer ears, and the like; they are all developments of old things, familiar to the anatomist in low, fishy forms.

ADDENDUM TO LECTURE V.

Here, again, I may remark that it does not enter into my plan to give an exhaustive bibliography, whether zoological, anatomical, or palæontological, but merely to set down the titles of such works as have been most useful to me in my special line of research, and which, therefore, may be of use to the reader.

With regard to the fossil types that suggest so much as to the development of the existing Mammalia, of which I have spoken in this fifth lecture, it seemed to me that it would be worth while to give a list of some of the papers, memoirs, and larger works that have come to hand during the last ten or a dozen years.

The Catalogue of the Fossils in the Hunterian Museum belongs to an older period; but it is very valuable, for it contains Professor Owen's description (with splendid plates) of the extinct *Glyptodon*.

BIBLIOGRAPHICAL LIST.

BETTANY, G. T. Esq., M.A., B.Sc., "On the Genus Meryochœrus (Family *Oreodontidæ*), with Descriptions of Two New Species." *Quart. Jour. of Geol. Soc.*, London, Aug. 1876, pp. 259–273, plates 17–18.

COPE, Professor E. D., "On the Extinct Vertebrata of the Eocene of Wyoming, observed by the Expedition of 1872, with Notes on the Geology," *U.S. Geol. Survey*, 1872, pp. 546–612.

———— "On the Flat-clawed Carnivora of the Eocene of Wyoming." Read before the American Philosophical Society, April 4, 1873, pp. 1–12, plates i.–ii.

———— "On the Short-footed Ungulata of the Eocene of Wyoming." Read before the American Philosophical Society, Feb. 21, 1873, pp. 1–37, plates i–iv.

———— "On the Primitive Types of the Orders of Mammalia Educabilia." Read before the American Philosophical Society, April 18, 1873, pp. 1–8.

COPE, Professor E. D., "Report on the Stratigraphy and Pliocene Vertebrate Palæontology of Northern Colorado," *Bull. U. S. Geol. and Geog. Survey of the Territory*, No. 1, pp. 1873, 9–28.

——— "On the Homologies and Origin of the Types of Molar Teeth of the Mammalia Educabilia," Philadelphia, March, 1874.

——— "Report on the Extinct Vertebrata obtained in New Mexico by Parties of the Expedition of 1874," *U. S. Geol. Survey*, part ii., vol. iv., Palæontology, 1877.

——— "On the Brain of Coryphodon." Read before the American Philosophical Society, March 16, 1877, pp. 616–620, plates i.–ii.

——— "On the Brain of *Procamelus occidentalis*." Read before the American Philosophical Society, May 4, 1877, pp. 49–52, plate i.

——— "Second Contribution to a Knowledge of the Miocene Fauna of Oregon." Read before the American Philosophical Society, Dec. 5, 1879, pp. 1–7.

——— "The Relations of the Horizons of Extinct Vertebrata of Europe and North America," *Bull. U. S. Geol. Survey*, vol. v., No. 7, pp. 33–54, 1879.

——— "Observations on the Fauna of the Miocene Tertiaries of Oregon," *Ibid.*, vol. v. part i., pp. 55–69, 1879.

——— "The Genealogy of the American Rhinoceroses." *Amer. Naturalist*, October 1880, pp. 610, 611.

——— "On the Genera of the Creodonta." Read before the American Philosophical Society, July 16, 1880.

——— "On the Foramina Perforating the Posterior Part of the Squamosal Bone of the Mammalia." Read before the American Philosophical Society, Feb. 16, 1880.

FLOWER, Professor W. H., F.R.S., "The Extinct Animals of North America," a Lecture delivered at the Royal Institution of Great Britain, Friday, March 10, 1876.

——— art. "Mammalia," *Encyc. Brit.*, vol. xv., 9th edit., 1883.

GILL, THEODORE, M.D., Ph.D., "Synoptical Tables of Characters of the Subdivision of Mammals, with a Catalogue of the Genera," Washington, 1871.

GILL, THEODORE, M.D., Ph. D., "Arrangement of the Families of Mammals," *Smithsonian Miscellaneous Collections*, Washington, Nov. 1872, pp. 1–41.

——— "On the Genetic Relations of the Cetaceans and the Methods involved in Discovery," *The American Naturalist*, vol. vii., Jan. 1873, pp. 1–11.

HUXLEY, T. H., F.R.S., "On a New Species of Macrauchenia (*M. boliviensis*), *Proc. Geol. Soc.*, 1860, pp. 73–84, plate vi.

LEIDY, Dr JOSEPH, "Contributions to the Extinct Vertebrate Fauna of the Western Territories," *Report of the U. S. Geol. Survey of the Territories*, vol. i., "Fossil Vertebrates," Washington, 1873.

——— "The Extinct Mammalian Fauna of Dakota and Nebraska," *Jour. Acad. of Nat. Sci. of Philadelphia*, vol. vii., 2nd series, Philadelphia, 1869.

——— "Description of Vertebrate Remains, chiefly from the Phosphate Beds of South Carolina," *Jour. Acad. of Nat. Sci. of Philadelphia*, vol. viii., Philadelphia, 1877.

MARSH, Professor O. C., "On the Structure and Affinities of the Brontotheridæ," *Amer. Jour. of Science and Art*, vol. vii., Jan. 7, 1874, pp. 1–8, plates i., ii.

——— "Fossil Horses in America," *American Naturalist*, vol. viii., May 1874, pp. 288–294.

——— "Introduction and Succession of Vertebrate Life in America," Address delivered before the American Association for the Advancement of Science, at Nashville, Tenn., Aug. 30, 1877.

——— "Principal Characters of the Coryphodontidæ," *Amer. Jour. of Science and Art*, vol. xiv., July 1877, pp. 81–85, plate iv.

——— "History and Methods of Palæontological Discovery," an Address delivered before the American Association for the Advancement of Science, at Saratoga, N. Y., Aug. 28, 1879.

——— "Notes of a New Jurassic Mammal," *Amer. Jour. of Science and Art*, vol. xviii., July 1879, pp. 1, 2.

——— "Notice of Jurassic Mammals representing Two New Orders," *Amer. Jour. of Science*, vol. xx., September 1880, pp. 235–239, figs. 1–2.

In this paper Professor Marsh remarks (p. 228):—"Mesozoic mammals have been very generally referred to the *Marsupialia*. An examination of all the known remains of

Mesozoic Mammalia, now representing upwards of sixty individuals, has convinced the writer that they cannot be satisfactorily placed in any of the present Orders. This appears to be equally true of the European forms which the writer has had the opportunity of examining. With a few possible exceptions, the Mesozoic mammals best preserved were manifestly low generalised forms, without any distinctive Marsupial characters. Not a few of them show features that point more directly to Insectivores, and present evidence, based on specimens alone, would transfer them to the latter group, if they are to be retained in any modern Order. This, however, has not yet been systematically attempted, and the known facts are against it. In view of this uncertainty, it seems more in accordance with the present state of science, to recognise the importance of the generalised characters of these early Mammals as at least of ordinal value, rather than attempt to measure them by specialised features of modern types, with which they have little real affinity." The original paper, however, will be referred to by the palæontological reader, not without feelings of profound gratitude to Dr Marsh and his fellow-workers for their excellent labours in this field.

OWEN, RICHARD, F.R.S., "Catalogue of Fossil Mammalia and Aves, contained in the Museum of the Royal College of Surgeons," London 1845.

SCOTT, Professor W. B. and Magee, W. F., "Preliminary Report upon the Princeton Scientific Expedition of 1882," Princeton, N.Y., 1882.

——— "Contributions from the E. M. Museum of Geology and Archæology of Princeton College," Bulletin No. 3, Princeton, N. Y., May 1883.

SUESS, E. VON, "Neue Reste von Squalodon aus Linz," *Jahrbuch d. K.K. Geolog. Reichsanstalt*, 1868, Bd. xviii. pp. 287–290, Taf. x.

LECTURE VI.

INSECTIVORA—*continued.*

My next instance is the Mole, a distant relative of the Hedgehog, and the head of another family. In some things the Mole is more remarkably specialised than the Hedgehog; he is less typical, and yet in him I find some of the most primitive mammalian characters. The lowest mammal living, but one, is the *Echidna;* I confidently expect to find a great correspondence between the skull of its embryo and that of the Mole. I shall show in this lecture how marvellously Marsupial some of the stages of the Mole's skull are; and I am rather inclined to speculate a little upon the retention, for a time, during growth, of such archaic characters in this old-world type. Great as is my goodwill towards the Mole, I do, nevertheless, look upon him as a coward. The Hedgehog has more pluck in him, but he defends himself by a spiny skin, as some ungracious people do by a spiny temper. When the old representatives of this most ancient family found their hunting grounds invaded by the higher Eutheria—Badgers, Stoats, Cats, —*et hoc genus omne*—who came to hunt them, they betook themselves to the lower parts of the earth.

There Nemesis followed them; they became sandblind, for the eyes not being used, ceased to grow after the time of birth. The thought of the Mole, when safety was assured to him, was to revel in the enjoyments of the table, and of all the fat and greasy citizens of the field or forest, he is the fattest and the greediest; his appetite is like that of Homer's heroes—"a rage of hunger." This sleek artful little hermit has slunk away from the outer active world for the sake of safety and creature comfort; not for moral improvement, for his temper is demoniacal. But if his temper is not perfect, his structure is; and if a second Paley should arise to preach to us on natural theology, the Mole in his garden would be the only text he would need to take. The Mole, moreover, is an old friend of mine, and much admired by me; long ago, we spent our days in the same field; I above, he below; but I used my eyes, he neglected his; he is still underground, but my eyes still see the flocks, and herds, and human face divine. Yet the Mole was not always a blind prisoner, making elegant inner prisons with his own hands; he is the modified descendant of some ancient full-eyed type.

And it was not the fault of any one particular parent, much less of the present Mole himself, that he was born blind; the degradation arose gradually, through the timidity of his progenitors, whose cowardice became hereditary and intensified; nature, however, avenged herself on his family—*she* never forgives a transgression.

Nevertheless, she loves and rewards workers; and this small, blind Sampson does, in his prison, feats that would seem fine sport to any one observing them. In softish earth he almost swims, whilst with hands and feet he pursues his way, and sinks, or creeps, or wades, or flies, through the dark, heavy medium. Perfect are the instruments that nature has given him for his digging work, and perfect are the instruments she has given this little waster, wherewith to destroy the countless living creatures of lower sorts that are struggling in the dark, for life, with him. The Mole did not become subterrestrial just lately, but a long time ago; his hereditary digging-paws show themselves, as diagnostics, when the embryo is one-third of an inch in length. The early appearance of the forwardly-placed, huge, digging hands, is surely evidence of an old inheritance.

The power of hearing in the Mole evidently compensates, in some degree, for want of sight; the concha is aborted, but there is a jointed cartilaginous meatus (or tube), and a very perfect and labyrinthic ear-drum. So that Caliban spoke well when he said—

"Tread softly, that the blind Mole hear not your foot-fall."

I know of no creature the study of which would so well reward the teleologist as the Mole; and none that would lead him into such bogs of difficulty, if he were a mere teleologist. For, nature, or the morphological force working within, produces curious structures, of various kinds, that are of no use to him, and that, indeed, are not left there to be of any use; these scaffoldings represent such buildings as were useful in the olden time. Here, if a man leans on teleology, alone, it will break, and pierce his hand; if the development of the Mole does not teach Darwinism, there is no use in observation, or in making deductions from observed facts; the morphologist's occupation is gone. The early development of the germ and embryo of the Mole has been largely worked out of late by competent embryologists, for it is a type easily obtained.

My task begins where theirs ends, and I follow the growth of the frame when the main work of differentiation is over;—when there has been formed cartilage for things of cartilage, muscle for things of muscle, and bone for things of bone. Ten stages have been followed in the head, point by point, part by part.

When we build, we prepare our work without, and make it fit for ourselves in the field, and afterwards build our house. For, having no true creative power, we have no option; we allocate, but cannot differentiate; our creations are mere toys, after all, for the great earth-mother has given us the prepared materials to play with. She, however, works in a fashion that infinitely transcends ours, and grows all her varied materials out of an apparently simple stuff, a pellet of which, no larger than a mote in a sunbeam, is enough for her to start a Whale with—a Whale that shall be mother of all the Whales that may come in that line. The embryo of that small sapper and miner, the Mole, may be traced through a dozen stages, each of which is the temporary equivalent of what might have been the permanent form of some type of Vertebrate. That is, these may be traced, as answering to so many platforms or levels of life; none of these, however, in this case, is fitted for free individual existence except the last; the others are passed through, not stopped at. Yet in the lower types of the Vertebrata we know that steps and stages in the life-history of a type may be fitted for free existence, by the development of temporary organs; in these we have true metamorphosis. But the rapid change of parts that are of no use to the individual is of the greatest interest to the biologist, who sees in them the writings of an old history, the records of lost peoples. If this be the true interpretation of these changes, and if, as is very probable, the Mammalia existed, even in the times that gave us the Primary fossiliferous rocks, species, genera, families, and orders of Mammalia may have come in, and gone out, in countless numbers, before the appearance of the types with which we are now familiar.

My work at the Mole begins with an embryo, which, in its curled-up form, is scarcely larger than a mustard seed (see fig. 1, p. 15). In this stage the various organs are developed, in rudiment at least, and the general form is quite similar to that of the embryo of any known Vertebrate, except that the post-oral arches are fewer in number than in fishes. Nor would a thorough embryological investigation as to the conditions of things at this stage, belie the outward form, since, part for part, and organ for organ, the oneness of this with that of any other type would be found to be, if not complete, yet very great.

In my next stage, when the embryo is the size of a

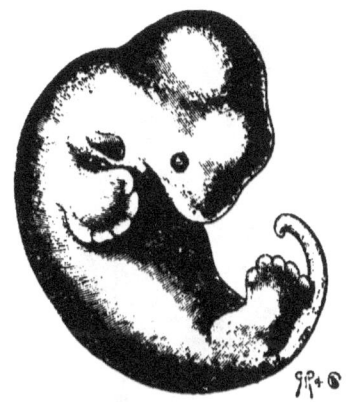

Fig. 12.—Embryo of Mole (*Talpa europœa*), gnified 7 diameters (2nd stage).

little nut-grub, that is, one-third of an inch long, measured along its curve, the form is very rapidly specialised, and remarkably so, the fore-limbs particularly; for although these are only, at present, broad flippers, with marks of the five toes upon them, yet their peculiar position, close to the head, reveals the type at once. Yet, although the hereditary agriculturist is even now revealed, the *connective-tissues* are all in a generalised state; cartilage is promised, but it is only marked by the greater crowding of the cells, and not by any definite modification, as yet. Where general embryology ends, there special morphology begins. My proper cue is the formation of clear or hyaline cartilage; the true primary skeletal material. We get that by the time the embryo Mole is the size of a Blue-fly's larva; this

Fig. 13.—Embryo of Mole (*Talpa europœa*), magnified 4 diameters (3d stage).

is the first of ten stages in which I have followed the details of the anatomy of the head.

I do not say *skull*, merely; given a dry skull, prepared by an assistant, and an artist to make the drawings for you, and you have a short and ready method for becoming a comparative craniologist. But that is not enough for this generation of workers. We mean by the expression "morphology of the skull" infinitely more than that. I shall merely mention certain things that have turned up in the course of several months' steady work at these types; especially showing how far the Mole agrees with, or differs from, the Hedgehog, in its development. In the earliest stages, these two kinds are very similar in many respects, but some of the most remarkable modifications of the part which becomes the malleus take place in young Moles before they leave the nest. There is a cartilaginous pterygoid rudiment, as in the Hedgehog, and the basis-cranii is

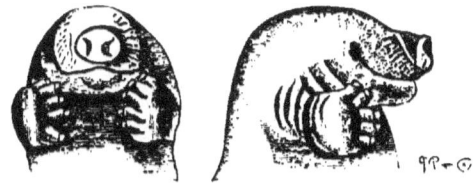

FIG. 14.—Embryo of Mole (*Talpa europæa*), two views of head, magnified 2 diameters (4th stage).

pneumatic as in that type; air-cavities like those seen in Crocodiles and Birds being developed there. There is an additional bony centre found in the manubrium or handle of the hammer, besides the internal deposit of bone formed in its head. But the perichondrial (or outer) bone, answering to the external articular tract in the lower jaw of the oviparous types, soon takes on a very remarkable form; it becomes *wild-grown*, so to speak, and three sub-distinct parts can be traced. These are seen to answer to the articulare externum, the supra-angulare, and the angulare of the ovipara. More than this, a sickle of bone passes across in front of the tympanic cavity, exactly as in the Marsupials. But this pretympanic bar is soon absorbed, and so, after a time, is the rest of the wild growth of fibrous bone, until, in the adult, the processus gracilis, or slender process of the hammer bone, is a mere point. In the Crocodile the cavum tympani (drum cavity) is mainly formed by the hollowing out of the quadratum (or jaw-pier). In the Bird it is behind that bone, which, however, is hollow, and opens into the cavity by a considerable passage. In the Mole, the incus, malleus, and stapes (the small ear-bones), lie in the drum cavity, and like their counterparts in the Bird—the quadratum, articular bone and columella (stapes)—are all pneumatic; little, hollow shells of bone, opening into the general cavity. In some adult Insectivora the sheath of the stapedial artery, which runs from the common carotid to the artery of the lower jaw, becomes ossified, and as this passes through the foot-hole of the stirrup, that bone is fastened in its place, the little bony tube being continuous with the os petrosum (or stony bone of the inner ear) at each end. This curious state of things is only temporary in the Mole; in young specimens, three parts grown, it is seen, but this tubular rod becomes absorbed afterwards. As in Birds and Crocodiles, the adult Mole's skull is extremely pneumatic, or full of air cavities. We only retain the mastoid cells in addition to the typanic cavity; these are in the hard lump of bone behind our outer ear.

The parts of the hind skull, especially, are well anchylosed or melted into a common mass, as in Birds and Bats, not, however, for the same teleological reason, manifestly.

The whole skull is a wedge, and the nose a borer, fastened to the sharp end of the wedge.

One thing surprises the student of the high forms of mammals, namely, that in front of the petrous bone (or stony ear capsule) in the Mole, there are two temporal scale-bones, apparently. The one in front of and below the other is the true superficial squamosal, or *squama temporis*; the one behind and above is a large semi-oval tract of the original cartilaginous skull wall, ossified by the prootic bone (petrosum). The temporo-mastoid bone (opisthotic), which in us forms the lump just spoken of, ossifies the cochlea (or snail-shell), the mastoid region, and most of the walls of the capsule where the semicircular canals, most elegant parts of the labyrinth, are embedded. The occipital arch, or back of the head, is very large and swelling, and the condyles or hinges for the first neck-joint are large also; the jugal arch (cheek) is perfect, but very delicate.

The hyoid arch—the second, or that which carries the tongue—is complete, but it is swung, as in Mammals generally, and also in Frogs and Toads (which curiously anticipate many things in mammalian morphology), from the skull, under the ear; the uppermost piece, or pharyngo-hyal, being set free for auditory purposes, and becoming the stirrup-bone.

Supposing the theory of the slow secular transformation of the old general types, into new special types, to be true, then the existing Mole, in its perfection of adaptive structure, has been as long in coming to its present perfection as the larger and nobler prone, or erect, types that trample the earth over its head. In its own line, doing its own dark work, it is as complete a creature as the clear-eyed, super-terrestrial types; as a Mole it is consummate—a complete and perfect example of a subterranean tyrant. In its turn, this cowardly creature tyrannises over other earth-borers, and very cunning is he; there is but a step from his stall to his larder; all around him are hosts of juicy grubs and worms, "and thereout sucks he no small advantage." Concerning tastes there is no disputing; one naturalist is fond of Whales; another of Moles, Shrews, and Mice. All these amusing types must have their supply of food; the great mother, Nature, loves all, and shakes out of her lap plenty for every kind. When we reflect that our own country possesses about 1200 species of insects,[1] and that some of the species are prolific beyond all calculation, then we come to understand how the higher Insectivorous tribes—Birds or small Mammals—find so plentiful a table in the wilderness.

[1] So Mr Stainton informs me.

The hungry, impatient Cat, who mistakes a Shrew for a Mouse, and then leaves her musky prey untasted, would starve upon that which fattens the Mole, the Shrew, and the Bat. The last of these kinds hawks for his small prey; but the Shrew, with his delicate proboscis, his sharp eyes, and his quick ears, knows where small beetles most do congregate. These he crunches and munches with exquisite teeth, the cusps or points of which are of a deep ferruginous red colour; more beautiful, strange to say, because they are thus stained.

The Power that made the beetle strong in his polished and enamelled armour, made also the teeth of the Shrew most fit instruments for crushing that armour in which the beetle trusts. It is pleasanter to look upon this vacillation, so to speak, of beneficent purpose, from the standpoint of a Darwin, than from the standpoint of a Paley; there is much that is painfully mysterious in the whole matter, and we only see it in a partial view. Anyhow, the human Vertebrate, who suffers from many kinds of hunger, is not better fitted for his mode of life than the Shrew for his; the latter is a perfect little creature, and well worthy the attention of the biologist.

We have three native forms of this widely-distributed family of the Insectivora, a family which has representatives in many parts of the old world, and in the northern part of the new.

The commonest kind (*Sorex vulgaris*) is intermediate in size between the other two; the largest is the Water Shrew

(*Crossopus fodiens*); the smallest (*Sorex pygmæus*) is not only *nearly* the smallest of our native Mammalia, it is almost the smallest of all the creatures that give suck to their young ones. Its rivals in smallness are the beautiful squirrel-like Harvest Mouse, who builds her grassy nest among the multiplied (or tillered) stems of a wheat-plant; and the least of our native Bats, the Pipistrelle.

There are not only appreciable differences between the skulls of the Land and Water Shrew, but in that of the former I find varietal modifications. The shoulder-girdle and breast-bone of these types is very instructive; this I have shown long ago; there is a considerable sternal, distal rudiment of the huge *coracoid bone* of the Monotreme attached to the manubrium sterni (or top of the breast-bone).

Most of the elongation of the snout takes place after birth; nestlings of about the tenth, or twelfth, day have a very proboscidean, or tapiroid appearance; they are like little Tapirs, with a long tail and short legs, or rather they are like Cuvier's fossil *Anoplotherium*. This middle stage is very useful to the morphologist, for at that time the various elements of the skull are easily traceable, the bones being separated from each other by tracts of cartilage, or membrane, as the case may be. These small Insectivores agree with the Ant-eater and Pangolin in having no cheek-bone (malar or jugal); in this they also agree with the Insectivora of Madagascar (*Centetidæ*)

—the Tenrec and his relatives, as I shall soon show. The occipital region or back of the head, and the interparietal and parietal roof-bones are all large; but the interorbital region is covered in by frontals, no larger, relatively, than those of a Snake; the small lachrymals (tear-duct bones) soon lose their distinctness, as do the bones of all the fore part of the head. The space between the base of the skull and the ear-bones is very large, and instead of *foramina lacera,* or ragged interspaces between the skull, proper, and the petro-mastoid bones, we have a considerable space, right and left, merely membranous. Above and behind the long narrow squamosal there is a bony tract, as in the Mole, which is ossified by the petrosum, (stony bone, prootic); this is the cartilaginous skull wall turned into bone, as in the Mole. The hard palate is well formed, and the pterygoid hooks for the muscles of the soft palate, are very small; the body of the pterygoid bone is swollen and cellular, but the basis-cranii has no caves, or recesses, to increase the size of the ear-drum, as in the Hedgehog and Mole. The lower jaw is a peculiarly elegant structure, and articulates with the squamosal, or temporal bone, in the glenoid fossa (hollow articular space), by two facets—an upper and a lower. The angular process of the mandible is rod-shaped, and is curved inwards at the end, as in Marsupials. In so small a creature as this, the ossicula auditûs, or small bones of the middle ear, are relatively larger than in the Mole, although really very small; they are quite normal, and the stapes (stirrup) is well shaped, with a large foot-hole; the rest of the hyoid arch is complete. The tympanic annulus, or ring-bone of the ear-drum, is rather slender, but has a broad flange on its hinder crus or limb; there is only a low ridge on the basi-sphenoid (skull-balk), helping to wall in the cavum tympani, or drum cavity; the concha or free part of the ear is well developed. Thus we miss in this diminutive type the cheek-bones (malars or jugals), and the tympanic wings of the basi-sphenoid; these parts have gone in the general reduction of the type; they have become small by degrees, and beautifully less, through the secular periods during which the Eutheria have been struggling upwards.

The ripe embryo, and the embryo when five-sixths ripe, are like little generalised Pigs, but their snout is obtuse, not discoidal, as in those rooting beasts. In their general appearance both the embryos and the nestlings might be taken as models for the restoration of the forms whose skeletons are found in the Upper Eocene deposits; only in size do they come short; some of those lost types (*Palæotherium, Anoplotherium*) were cattle of considerable dimensions. The probability is that the Shrews have, on the whole, kept most of their original characters as early Eutheria; but that they have undergone, besides gradual lessening of bulk, gentle changes of structure, in conformity with their habits. Their safety during all this time has been due to the fact that they have made themselves almost invisible, through their minuteness and their crepuscular habits—they are evening feeders, evidently, else how does the Owl manage to find so many of them?

We have, however, not only to consider the small size of the Shrews, but also their wide geographical distribution. The Shrews proper (*Soricidæ*) have as wide a range as most types, being found in North America and all over the Old Word. Now I am not aware that any members of this group ever travel any distance from their birth-place; for the most part, undoubtedly, they make but a short range in search after food. Accepting this as truth, how then shall we account for the fact, that although not nomadic in their habits, they are to be found nearly all over the habitable globe? Not reckoning a large number of related types, but merely considering the true Shrews (*Soricidæ*), we are brought face to face with the striking fact of their presence in so many places, and these, so widely apart. I say *true Shrews* purposely; and there can be no difficulty in supposing that genera and species differing so little from each other, whose variations of size, structure, and habits are so slight and gentle, may have arisen—most probably did arise—from one common stock. Let anyone think how long it would take, not merely to develop such a number of closely related species and genera, but also to apportion them their lots in life. The whole world was before the primary

L

Shrews, but it must have taken them more time than we can conceive of to settle down in their colonies, and to people the greater part of the globe with their pigmy tribes. This is not brought forward as though it were an insuperable difficulty—as a problem that cannot be accounted for on evolutional principles. What was wanted was an abundance of time. The forefathers of the Shrews were certainly early Eutheria that became modified mainly by dwarfing, and so did not develop upwards into nobler types.

Amongst other kinds of early Eutheria an opposite kind of variation took place. They became larger and larger, and according to the difference of their feeding-grounds, these being sometimes flat or hilly, soft or hard; on burning sands, or on frozen snow; their feet became modified, the innermost digit going first. The tooth-pattern, also, grew more and more complex, as the diet became in many cases of a purely vegetable kind. Then a purely vegetable diet is of necessity very varied; browsing is added to grazing; and thus the molar teeth have to be very large, with large bones to hold them, and the grinding crowns must be very effective millstones.

I may here give an instance of the freedom of selection as to the kind of food eaten by the most purely herbivorous type. The Horse, every one knows, is a most particular creature as to his diet; his sensitiveness to the odours of various kinds of food, especially, is very great. Some three-year-old Colts were turned for the winter into an excellent pasture, with the kind of grass they prefer,—close and velvety,—for they have a most accurate bite, and are not like the Cow, who will eat long, coarse stuff, its length helping her, she having no upper incisors. In the quick hedges bounding this pasture, the pollard ash-trees had recently been beheaded for poles, and the "top and lop" had been made into faggots. These trees are very bitter, and full of potash salts, besides; nevertheless, the Colts preferred this woody diet to the grass, and made a clean riddance of the firewood, only leaving the white centre of the largest boughs. The Rhinoceros is said to tear a young tree into laths, and then to eat the laths as we should eat celery. But the magnificent dentition of the Cow, the Horse, and the Rhinoceros has been preparing; —this is no fable; the palæontologist knows if this be true or not—ever since the time when the cattle were none of them any better, and many not so good, as the existing Tapir. It is not true that nature "does not make her works for man to mend." I have been familiar, from my childhood, with the method by which the breeds of the existing cattle are mended by us. During the long Tertiary epoch, during which nature, herself, has been developing the teeth and jaws of the improving Eutheria, she has also, at the same time, wondrously perfected their gait and carriage. No movement of a dancer is more elegant than the ordinary walk of a high-bred

Horse; see him in contrast with that sprawling quadruped, the Crocodile! The gait of the Insectivora is not all that can be wished, although their legs are fairly set under them; it is only when the whole weight comes to be carried by limbs ending, below, in a cloven or a solid hoof, that the last degree of quadrupedal perfection is obtained.

Go again to the Tapir, and see his paces—slower or quicker—if you will truly value the forms and qualities of a Stag, an Antelope, or a Horse. That Rome should have been made in a day would have been a mere game of toys, and not a miracle, in comparison with the sudden transformation of an Eocene Palæotherian into a modern race-horse. I may be asked, What has all this to do with the Insectivora?—Very much: the Insectivora have been preserved to us to show what simple creatures the early Eutheria were, and how little they are, now, an improvement upon the Marsupial types. The Tapir, also, in both the old and new world, is a living witness of what general forms our special Mammals must have arisen from.

An old Naturalist (*Scheuchzer*) more than a century and a half ago found the bones of a large Salamander, and took them for the remains of a Man who had seen Noah's flood; *homo diluvii testis*. That occurrence belongs to the recent things of the earth; the Tapiroids and quasi-Insectivores belong to the truly old things. Our Insectivores, and the Tapir, are the waifs or strays from almost extinct groups that could not make their *one* talent *ten*.

We are unable, without strong effort, to imagine what may be done by slow changes, through countless thousands of years; moreover, nature has not left herself without a witness as to what she can accomplish, rapidly, at times, when changes come suddenly upon living creatures. I am of opinion that too much has been made of slow, almost infinitesimal changes, taking place during long secular periods; nature has, in the morphological force, in relation to various surroundings, an unused surplusage of modifying energy that is practically infinite. Hitherto, metamorphosis has been mainly known in the Arthropods—insects and their relations— but we are waking up to the importance of this branch of our subject in the study of the Vertebrata; there are creatures, and not a few of them, who show us what the constructive energies of a living creature can do in a single life term. Everywhere, also, in the development from the embryo of each high type, we see the footprints or marks of the most marvellous transformations that took place in the green youth-time of the earth.

LECTURE VII.

INSECTIVORA—(*concluded*).

As to the doctrine of the development of the higher from the lower forms of Eutheria, I have, as yet, kept back some of the best witnesses; these are all foreigners. And yet there is no need that we should go beyond the sea to fetch witnesses to the truth of this doctrine, nor that we should go into the depths of the earth; the proofs are nigh unto us, and surround us everywhere. The Order Insectivora, in its present state, contains a great variety of types, and neither my time, nor my materials for work have enabled me to study much more than the representatives of about half the known types. In my last two Lectures, I have spoken of the structure of the skull, especially of the Hedgehog, the Mole, and the Shrew, each the head of a family, and famous amongst the tribes of the Insectivora. But I have worked out the skull in its early stages in *Centetes*, or the Tenrec; in *Galeopithecus*, or the Flying Cat; and in *Rhynchocyon*, a large, long-snouted, *proboscidean* Insectivore, from the east coast of Africa near Zanzibar.

I have only seen the adult *Tupaia*; of this type I

hope to obtain the embryos or at least the young, as it is one of the most instructive of these mixed or generalised forms. Of the Centetidae or Tenrec family the largest is the Tenrec, proper; it is larger than the Hedgehog, and has scarcely any tail; but, like the aquatic Whale, its head is about one-third the length of its body. It is armed with spines, and during the wet and cold months it sleeps, as the Hedgehog does here; those months, corresponding in time to our summer, as Mr Dallas tells us,[1] are the wintry months of Madagascar. Besides this kind, of which I have been able to work out several stages, I have studied the skull in its sub-adult stage in three of the lesser kinds of Centetidae, namely, *Ericulus*, *Hemicentetes*, and *Microgale*; the latter is as small as our common Shrew. Every one is aware that the straits of Mozambique separate Faunae and Florae that seem to bear very little relation to each other; the South Eastern African types, and the Mascarene types, whether of plants or animals, are very diverse from each other.

These Mascarene Insectivora are quite unlike the forms found at or near Zanzibar, and in the South African region generally; there the Elephant-Shrews, rightly named, and the Rhynchocyon, their relation, are extremely unlike the Tenrecs, and are indeed a sort of half Opossum, with a proboscis. The Tenrecs agree with the Hedgehogs in having a pneumatic skull-base, and with

[1] *Cassell's Natural History*, No. 12, p. 360.

the Shrews in having no cheek bones. When the development of these long-snouted types is studied, we find that the main skull beam in front of the pituitary body and its shallow cup, the sella turcica, is mainly composed of one large solid bar of cartilage just as in such diverse types, at each end of the Vertebrata, as the Skate and the Whale. Up to the olfactory region (middle ethmoidal territory) this single solid bar is flanked by a shorter bar, right and left, and these two bars are a direct continuation of the cartilage, which invests the cranial notochord—the "investing mass" of Rathke (the continuation in the head of the axis of the creature). The shorter paired cartilages were called by him the trabeculæ cranii or rafters of the skull; the middle projecting bar I have simply called the intertrabecula; Rathke's term has become classical. In the Skate the roofs of the nasal organs lie back against the front of the skull, but in this, as in most Mammals, the whole nasal roof is continued to the end of the snout, and is folded over to form the alæ-nasi, the cartilages that encircle the outer nostrils. Thus the long rostral median cartilage does not project as a free rod, as in the Skate, but its crested upper edge is everywhere confluent with the right and left cartilaginous roof-plates of the labyrinth of the nose. Therefore, from the outer nostrils to the naso-palatine openings, over the soft palate, there are two tubular passages; over these lie the turbinals, or coiled bones, on which, in the region nearest the skull, the olfactory nerves or nerves of smell are distributed. These scrolls in us are divided into the upper and middle turbinated bones; in front and below we have the maxillary or inferior turbinal, whose nervous supply is from the first or ophthalmic branch of the fifth nerve—the trigeminal—the nerve which supplies most of the upper and lower face, and this first branch is that nerve which sets up sneezing. Under the median bar or thick base of the septum nasi or partition wall of the nose-passages, we have a vomer or ploughshare bone, large and long in proportion to the great intertrabecular beam which it undersplices. There are also other little vomers, two in front protecting Jacobson's organs, and two behind, tying the legs of the main vomer to the turbinal masses, as in the Hedgehog. But that which, to me, is wholly unique, and morphologically inexplicable, at present, is the subdivision, by superficial peeling, of the main or middle vomer.

I have examined vomers of all sorts and sizes, in types varying from a Sturgeon to a Man; but this *barking*, as it may be called, of the median piece is to me something new. In the nearly ripe embryo of the Tenrec the bone is relatively very large, and is becoming rough, cracked, and cellular; but in young from the nest, about the size of Field Mice, the lower surface of the bone has split itself, like the bark of the plane-tree, into secondary flakes. There are two of these superficial centres of ossification, one before the other, and they sheath the inner and main piece, as it also sheathes the base of the nasal partition wall. I find this state of things in a sub-adult specimen of the lesser kind (*Hemicentetes*), and the smallest kind (*Microgale*); my specimen of the latter, *M. longicaudata*, has a tail nearly twice as long as its body, which is very curious for a member of the Tenrec family, the main type of which has merely a stump, and is termed *ecaudatus* on that account. This fact is worth remarking, for nature runs on her segments in the hind part of the axis of a Mammal in a seemingly hap-hazard manner; ready at any moment to "dock" them, as horse-dealers their "cattle," or as dog-fanciers curtail their pups.

I may remark, in passing, that I use the term "Nature" as a personification of the forces that are at work in organisms, feigning her to have a quasi-human character, —partly kind, partly cruel,—because to us Western people, here, in the end of the nineteenth century, the old Eastern custom of ascribing both evil and good directly to the DEITY seems irreverent and profane. Let that pass;—the biologist stops his ears for the time to the groans of this nether world, and goes on, with what cheerfulness he can muster, at his own proper work. He will be ready to argue about the origin of evil, and the meaning of pain, when he has the proper data, which will not be until the "consummation of all things."

Nature has been a good mother to the Tenrec; she has set him up in life with a good stock of all that he can want; and when he is weary she gives him several months of sleep, continuously. Also, she has defended him and all his family, making his workingday coat sharper than a thorn-hedge, so that he who would take him must himself be well fenced and armed. The Tenrec is one of the largest of the Insectivora, and his large bony skull is crested behind like that of a Tiger or Hyæna, so that his temporal muscles must be of great strength. Yet the action of the jaws cannot be quite like that of a Tiger or a Hyæna, as he has no cheek bones. The teeth are very strong and sharp-crested, so that the larger Beetles, and the lesser Vertebrata, are easily broken up by him; everything of this kind is grist to his strong mill. The ear-drums are like those of the Hedgehog, the tympanic bone being of a low and prototherian type; it is supplemented by laminæ or wings that grow from the base of the hinder sphenoidal bone, and thus the basis-cranii is excavated, as it were, as in Crocodiles or Birds. I consider this character to be the most typical of any as an Insectivorous diagnostic; and yet it is not constant. There is nothing constant in the order; it is a sort of low, tentative, transitional group, and a zoologist has to take certain characters, many of them in the soft parts, relating to the reproduction of the species, and make the best of them he can. In passing, I may remark that the Tenrec is more prolific than that most productive, oldfashioned, generalised herbivore—the Sow. I have heard of this creature having sixteen young at a birth; the

Tenrec has as many as eighteen or twenty. If that is not a sort of reptilian character, I do not know what is.

This family and others, especially the Lemurs, give the Mascarene fauna a very peculiar, and extremely interesting character; all the types, as I have said, are marvellously unlike those found on the mainland of Africa, beyond the straits of Mozambique. These things set both the biologist and the geologist speculating as to the changes in the earth's crust that have brought about these remarkable isolations of types, isolations largely due, undoubtedly, to extinction of species. At present the fragmentary state of our knowledge excludes all boasting, so that the biologist is kept in a state of chronic humility. Here, if anywhere, caution is necessary; we must not imitate the old geographers, who would not suffer any blanks to be left in their maps, and thus came under the lash of the satirist—

> "Geographers, in pathless downs,
> Put elephants instead of towns."

Some of the most remarkable and instructive of this variable Order are found just north of the equator and of Wallace's line in the south-eastern part of Asia and the contiguous islands. Of course, south of Wallace's line, in the Australian region, we have Marsupials, and coupled with them, the lowest of the living Birds. But some of the Eutheria living a little further north show that they are not many steps in advance of the low southern types. In Java, and other islands near, as well as on the main land, even as far as to India, there is a kind of Insectivore—the Tupaia, which is very isolated from the others in the Order. The species of this genus are elegant squirrel-like creatures, but they are rather related to the Lemurs than to the Rodents. I only know this type in the adult state, and I have only carefully studied the skull; but this is enough to make me long for a fuller acquaintance with it. There are two important things in which this skull differs from that of the Insectivora generally, namely, its complete orbital ring, and its large bulla tympani or additional bony shell to the drum-cavity. The orbital ring is complete in Lemurs, the Primates including Man, and in some Ungulates; but this is the lowest type of Mammal, I believe, in which this character is perfected. In the Tupaia, however, this ring is peculiar, for there is an open space where the zygomatic process of the squama temporis (temporal bone) rests upon the malar or cheek bone.

This type has, certainly, the most perfect and Cat-like bulla tympani; yet I suspect that in this case, as in a type soon to be mentioned—*Rhynchocyon*—it does not become cartilaginous before it is ossified: in the Cat this innermost ring of the meatus exists as hyaline cartilage for some time. This isolated group of Insectivora—the Tupaiadæ—comes in with excellent evidence as to the origin of the higher forms of the Eutheria: all these mixed or generalised types are the living, but modified, counterparts of the early existing Mammals. In the Tupaia we have, evidently, a type which varied in two directions at once; it became, in a sense, double-minded; it was, therefore, impossible for it to ascend to any height towards any one of our modern groups; its tendency towards the Lemurs hindered it from becoming a carnivorous beast.

FIG. 15.—New-born young of Flying Cat (*Galeopithecus*) ——? sp. two-thirds nat. size.

If the last type, the Tupaia, is an upward-looking form, the next to be mentioned—the Colugo, or Flying Cat (*Galeopithecus*)—certainly looks downwards.

This latter type is the opprobrium of zoologists, as such; it will not fit in well anywhere; but to evolutionists, it is a very god-send. The Insectivora have got it amongst them, at last, but it is as lonely as a stray Elephant in a pathless desert. This type is found in Borneo, Sumatra, and in the Philippine Islands, so that it is a near neighbour of the Tupaia family. Through the kindness of various friends, for whose help I am grateful, I have been able to work out the skull in the new-born (or ripe) young, and in two more advanced young taken in the woods, with their mother, in the Philippines, by Prof. Moseley. I have also studied the skull in the adult. I must confess that the skull, to say nothing of the rest of the creature, is to me a perfect puzzle, looking at it from a zoological standpoint; but I see nothing in it that I am not familiar with as a morphologist, except intense specialisation. Yet this is just one of those forms which, whilst lying at the very base of its own group, is specialised for its own life-purposes quite as much as the other members. The Sucking Fishes (Marsipobranchs), and the Serpents, are also instances of this kind. The types of Mammalia that are nearest to Man do, indeed, come much closer to their natural overseer or master, than any known Insectivore does to the Colugo or Flying Cat.

Everything that I have seen suggests, what Mr Wallace long since asserted, namely, that this is a most ancient type. It is very stupid: it won't die when you have done your best to kill it—like a Frog or a Snake;—if the Cat has nine lives, it has ninety-and-nine. It used to be thought to be a Lemur; then a Bat; it is almost equally akin to a flying Phalanger—a Marsupial; for, as shown in the woodcut, there is a very marked rudiment (so it appears to me) of the pouch to be seen in the embryo. I do not, indeed, find it very much like the existing flying Marsupial (*Petaurus*), but we have only odds and ends of the old Metatherian group left alive, and I fancy that this is an old flying Phalanger, that has just flitted, one may say, into the Eutherian territory. Yet the Bats certainly might claim relationship to him, as a sort of arrested forefather, so that it may be said of Nature that she tried her 'prentice hand on the Colugo, and then, afterwards, she made the Bats.

Nature must have broken up all her old types of Bats, and buried them; just forgetting to get rid of the very first, and poorest, of these—the Colugo. If this be the truth of the case, we have in this beast a scarcely modified, early Tertiary Bat; not a Bat in the modern zoological sense, but an Insectivore with the very rags and tatters of the Marsupial upon him, striving to make to himself wings, that he may scorn the earth. Instead of going into details with regard to the anatomy of this type, I will just mention two or three things in its structure that mark it out as one of the most unique and lonely forms in existence. Yet zoologists have not been so bold in this case as in that of the Hyrax, which is made to stand now as the head of the family Hyracoidea.

The feeling is that you cannot make a faggot of

one stick; yet other kinds must have existed; and the supposed and the actual, put together, might have served to make a new family, the "*Galeopithecidæ*."

The parachute of this beast is very perfect, more perfect than that of the flying Phalanger, or the flying Squirrel; the fold or skin runs over the dorsum of both hands and feet up to the nails, along the sides of the neck, and in front of the arms, and also along the sides of the tail, behind the legs and thighs.

Some Bats (*Desmodus*) have their front teeth slightly notched, but the Colugo, having only the side incisors, above, has the middle and next incisor, below, notched up to its "cingulum" into eight or nine toothlets—like a comb.

No other Mammal has such teeth as these. The molars and premolars are strong and large, and they are socketed in a very strong and solid jaw; it is a vegetable feeder. Between the upper jaws the hard-palate is very large, both in width and length, and is very complete; quite unlike the Metatherian palate, and that of many of the more typical Insectivores. The cheek bones, also, are large and strong, but the post-orbital region is open to the wide temporal region, which is a huge fossa for the large temporal muscles. In one respect the only skulls that come near that of the Colugo are those of the winged forms, both of the Sauropsida and Mammalia—Birds and Bats. In these skulls the landmarks are all gone, except the holes for

M

vessels and nerves, and the notches and ridges that are in all skulls; all the sutures are filled in with bony matter. Yet, whilst thus agreeing with the Bats in the complete ankylosis or union of the cranial elements, they have disappointed me in not yielding what I expected to find, namely, a correspondence, in form, with the skull of the "Pteropinæ," or frugivorous Bats. They come nearest in size to, and the largest of them are fellow-countrymen with, the Colugo, but skulls of both types placed side by side show only the ankylosis, and the general characters, in common. As they live a similar life, I expected to find the tympanum, or ear-drum, similar in both, but that cavity, in Bats, is developed much more in the fashion of that of the Tupaia, and of a kind, next to be described, from Zanzibar, and not like what is seen in the Colugo. In the lesser embryo of *Galeopithecus*, the annulus, or drum-ring, is a strong, thin, flattish dish of bone; afterwards, in older specimens, I find no bulla, or additional swollen part, either in bony or cartilaginous tracts; but one thing I do find, namely, that the bony tract soon creeps along the meatus externus, or outer passage; quite an unusual thing in an Insectivore;—in Mammals higher up, this is constantly seen. In *Galeopithecus* even this is done in a strange manner; for the squamous bone grows strongly downwards in front of the meatus, or passage, by its post-glenoid process, and behind that passage by the downward sang from its hinder part,—the great post-squamosal process. Thus in the dry skull of the adult we see the meatus or ear-passage as a thick-lipped chink, the small opening being vertical; its walls are the inwedged outgrowth of the annulus, or ear-ring. All these parts are so completely fused together that nothing but the fœtal or embryo skull will show how far each bony "centre" extended. There are no basi-sphenoidal excavations, or air cavities, in the base of the skull, as in the Hedgehog and Tenrec.

The ossicula auditûs, or small bones of the ear-chain, are formed and finished very early, but they show no leanings towards those of the Marsupials.

There is one thing, however, in this skull which is more important to the developmentalist than all the others put together. During the many years in which I have worked at this subject, there is one bone which I have come across, again and again, but always with renewed interest. The name of this bone—the "parasphenoid"—is not found in treatises on human anatomy. It is the first bone in the vertebrate series taken from the more superficial structures to support the deep cartilaginous endo-skeleton, and the first to appear in the embryos of Fishes and Amphibia. To one familiar with the Frog's skull, this is a striking part of that instructive piece of cranial architecture, which is intermediate between that of the lower and the higher tribes. Under the Frog's basis-cranii there is a dagger-shaped bone, this is the parasphenoid; it supports the cartilaginous skull, but is distinct from it: the blade underlies the orbital region; the guard the temporal region—right and left; and the haft the basi-occipital.

In air-breathing types, when the endo-cranium or inner skull becomes more and more solid, the cartilage becoming transformed into bone, the parasphenoid becomes inconstant. In Serpents there is only the blade, in Lizards a rudiment, only, of that part; in Crocodiles only a rudiment, right and left, of the guard; and in Chelonians (Turtles and Tortoises) none whatever. That it should be very large in all Birds, although broken up into three centres of ossification, is truly remarkable; for the reptile is but a *Pupa*, in respect of the Bird, which is an *Imago*, so to speak. He who would find out this riddle must employ Darwinian methods. The reptilian nature of Birds is doubted by no one capable of judging; but reptiles of the existing type are not their progenitors; and I question if any true Amniotic (or true Reptiles) ever were. Twenty years ago I showed that the temporal scale squamosal of the Ostrich family (flat-breasted Birds—Ratitiæ) was a truly amphibian bone, just like that of a Frog or Newt, and answering to both the "preopercular," and "temporal scale," of a Ganoid Fish, in one single bony centre. That fact, the large size of the parasphenoid, and other things that might be mentioned, being put together, show conclusively, that the flying Fowl came from a low, and most probably a gill-bearing, amphibian ancestry. Does such an amphibian character as this remarkable supporting splint-bone of the skull ever turn up in the Mammalian class? It is found in this very Flying Cat I am speaking about; it has a rudiment of the blade, exactly like that of a Lizard, and I know of no other Mammal with this most important rudiment. But, long ago, in the most stupid and ancient of the Rodents or gnawing animals, a South American type—the Guinea-Pig— I found large remnants, right and left, of the guard; and lesser rudiments of this part in the Rabbit, the Hyrax, and also in Man. Every student of human anatomy knows that, besides the various bony centres that are formed in the cartilaginous basis cranii under the pituitary body, there is a free, independent nucleus, right and left, in the angle formed by the base and wing of the posterior sphenoid. This is just where the os petrosum or stony bone of the ear-labyrinth is wedged in, and where the internal carotid arteries enter; these bones are called the "lingulæ sphenoidales," or little tongue-like processes of the sphenoid bone. Let it be conceded, then, that in the Mammalian skull, from that of the Flying Cat up to that of Man, remnants of the parasphenoid of Ganoids, Dipnoi (double-breathing Fishes), and Amphibians are found, and you open a door to the most daring Darwinian speculations ever promulgated.

The Rhynchocyon is a much more intelligible beast

than the Flying Cat; it has relations, near and more distant, and is indeed one of the most remarkable creatures, in this respect, of any among the Insectivora, having many affinities. This type is closely related to the Elephant Shrews of South Africa (*Macroscelidæ*), but is larger, being about the size of a Rat; the home of this form is near Zanzibar (South-East Africa), and is

Fig. 16.—Advanced Embryo of *Rhynchocyon cernei*, from Zanzibar, ¾ nat. size.

separated, therefore, from the Tenrec and his relations, merely by the Mozambique Channel. As I have already said, that Channel runs between a widely diverse Flora and Fauna, types that are very unrelated to each other. I am indebted to Dr Dobson for a nearly ripe embryo of this type; I have, at present, merely worked out the skull of this valued specimen, but it has rewarded and delighted me more than any kind I have received for a long time past. If nature had triturated together the germs of four or five types of Mammals, and had then made this mixture grow, she could scarcely have developed a more curious and composite creature than this long-nosed Insectivore. When Prof. Huxley propounded his oft-quoted theory of the evolution of the Mammalia, he might have known the structure and development of this type by inward light. Nothing of the kind, however, is ever revealed to biologists in that manner, we only get our facts by opening out the fine folds of organic forms with needle and scissors; we do unroll a good number of these small scrolls, but it is painful and patient work. I am satisfied that no searcher after the evidences of evolution ever saw anything more instructive than what I have found in this small beast.

I will make a catalogue of its characters, and then close this Lecture.

1. It is a proboscidean; the double, elongated nose-tube is jointed; there are thirty segments in the ali-nasal cartilage, on each side, and these jointed tubes are both united and divided by the long "inter-trabecula," or continuation of the septum nasi or partition wall of the nasal labyrinth.

2. The tympanic cavity is formed after the manner of that of a Cat—a large os bullæ (swollen bone) lying inside

the ordinary annulus (tympanic or proper bone of the drum).

3. Besides an anticipation of the carnivorous type of skull, there are plain and evident memorials of that of the Marsupial; for the ali-sphenoids each give off a trumpet-shaped process to enlarge the cavum tympani or cavity of the ear-drum; a most diagnostic character in the Metatheria.

4. I have already spoken of a very peculiar structure seen in many of the Australian Marsupials, namely, that the posterior paired vomers are large, for they meet behind the moderate-sized middle-piece; this is well seen in Rhynchocyon; there are several more characters equally marsupial.

5. The very diagnostic tympanic wings of the basi-sphenoid are absent in this peculiar Insectivore.

6. The pre-sphenoid or third basal beam-bone is distinct, as in Marsupials and Rodents.

Thus this greatly specialised kind of Insectivore, whilst retaining the most marked characteristics of the Metatherian skull, takes on two characters, one of which, had it become dominant, would have landed it amongst the Proboscidea or Elephants, whilst the other would have made it a Carnivore. It attempted too much at once, and thus, like a man in doubt, it made but little progress; moreover, in this developmental shilly-shallying, it failed to drop the old Marsupial, to take on the new Eutherian, nature, and was thus in danger of going out of being with many of the members of that much-extinguished Order. Other types, not thus confused in their ambition, worked out the old strain of Metatherian degradation, and taking to one definite line of ascent, put on new specialisations in harmony with their surroundings, and to this day their descendants are the rulers of the forest and the field.

ADDENDUM TO LECTURE VII.

Bibliographical References.

ALLMAN, Prof. GEORGE J., F.R.S., "On the Characters and Affinities of Potamogale," *Trans. Zool. Soc.*, vol. vi. plates i. ii., woodcuts figs 1–9, pp. 1–16.[1]

ALSTON, E. N., "On an Undescribed Shrew from Central America," *Proc. Zool. Soc.*, 1877, pp. 445, 446.

ANDERSON, JOHN, M.D., "On the Osteology and Dentition of Hylomys," *Trans. Zool. Soc.*, vol. viii. art. 13, plate lxiv. pp. 453–467 (1874).

AUSTEN, N. J., "On the Habits of the Water Shrew (*Crossopus fodiens*), *Proc. Zool. Soc.*, 1865, pp. 519–521.

BARBOZA DER BOCAGE, Dr J. V., "Sur quelques Mammifères peu connus, d'Afrique Occidentale, qui se trouvent au Muséum de Lisbonne," *Proc. Zool. Soc.*, 1865, pp. 401–404. (*Bayonia velox vel Potamogale velox.*)

BRANDT, J. F., "On Solenodon," *Memoirs of the Imperial Academy of St Petersburg*, 1832–3.

COUES, ELLIOT, "Precursory Notes on American Insectivorous Mammals with Description of New Species," *U. S. Survey*.

——— "On Scapanus Americanus," *American Naturalist*, No. 13, p. 189 (1879).

[1] In this paper, besides the excellent figures illustrating the anatomy of this, the noblest of the Insectivorous types, there is a splendid coloured plate (i.) from one of Wolf's originals. Every such plate is an invaluable addition to zoology.

DALLAS, W. S., "Insectivora," in *Cassell's Natural History*, part ii. pp. 342–348, and part iii. pp. 349–384.

DOBSON, G. E., M.A., F.R.S., "A Monograph of the Insectivora, Systematic and Anatomical." London, John Van Voorst. Part i., 1882, and part ii. 1883.

FLOWER, W. H., LL.D., F.R.S., Article "Mammalia," *Encyc. Brit.*, 9th edit. vol. xv. pp. 400–405.

GILL, Dr THEODORE, "Synopsis of the Insectivorous Mammals," *Bulletin of the Geological and Geographical Survey of the Territories*, No. 2, Second Series, Washington, May 14, 1875.

GRAY, J. E., F.R.S., "A Revision of the Species of Golden Moles (*Chrysochloris*), *Proc. Zool. Soc.*, 1865, pp. 678–680.

GÜNTHER, Dr ALBERT, F.R.S., "Description of a New Species of *Chrysochloris* from South Africa," *Proc. Zool. Soc.*, April 6, 1875, plate xliii. p. 311.

———— "Remarks on some Indian, and more especially Bornean Mammals," *Proc. Zool. Soc.*, May 16, 1876 (*Tupaia*, 1–9), plate xxxvi. pp. 424–427.

———— "Notes on some Japanese Mammalia," *Proc. Zool. Soc.*, June 1, 1880 (*Urotrichus* and *Talpa*), plate xlii. pp. 440, 441.

———— "Notes on the Species of *Rhynchocyon* and *Petrodromus*," *Proc. Zool. Soc.*, 1881, pp. 163, 164, plate xiv.

MIVART, ST GEORGE, F.R.S., "Notes on the Osteology of the Insectivora," *Journal of Anatomy and Physiology*, i., 1867, pp. 281–312; ii., 1868, pp. 117–154.

———— "On *Hemicentetes*, a new Genus of Insectivora, with some additional Remarks on the Osteology of that Order," *Proc. Zool. Soc.*, 1871, pp. 58–79.

PARKER, W. K., F.R.S., "Shoulder-girdle and Sternum," *Ray Soc.*, 1868, plates xxvii. xxviii. pp. 210–213.

PETERS, Dr W. L., "Neue Säugthiergattungen aus den Insectenfressern und Nagern," *Monatsber. Akad. der Wissensch.*, Berlin, 1846, pp. 257–259.

———— "Ueber die Säugthiergattung, *Solenodon*," *Abhandl. Akad. der Wissensch.*, Berlin, 1863, 1864, pp. 1–22, plates i.–iii.

———— "Ueber die Classification der Insectivora (besonders, *Ericulus, Echinogale*, und *Potamogale*)," *Monatsber. Akad. der Wissensch.*, Berlin, 1865, p. 286.

SUNDEVALL, C. J., "Obersight af Slaglet Erinaceus," *K. ret. Akad. Handlgr.*, Stockholm, 1841, pp. 215-240; Isis, 1845, pp. 273-280.

THOMAS, OLDFIELD, F.Z.S., "Description of a new Genus and two new Species of Insectivora from Madagascar," *Jour. of Linn. Soc.* (Zoology), vol. xvi. (Read March 2, 1882, pp. 319-322.)

LECTURE VIII.

THE REMAINING ORDERS OF MAMMALIA.

IN seeking to creep upwards towards Man's high place in nature, I need not necessarily pass in review all the Orders of the class to which he belongs. The members of several of the groups of the higher Eutheria are, in certain respects, much more specialised, or changed from the original five-toed types, than Man himself. But there are certain things in some of the lower Eutheria, not yet brought under review, that are very instructive; these things will now be spoken of. The Bats, Rodents, and Lemurs are manifestly the next in order above the Insectivores. The second of these groups (Mice, Bats, Squirrels, &c.) have less to interest us than the third, the Lemurs; yet they are all full of teaching as to the possible specialisation of a general Mammalian type.

Chiroptera (The Bats).

The Bats are evidently a modification of the Insectivorous type, and are, for the most part, themselves, Insectivorous. There are two types not belonging to the

Bats, and far apart from each other, that suggest to us how these long-fingered, leathern-winged folk came into being, namely, the Galeopithecus, already spoken of, and the Aye-Aye, a strange kind of Lemur. Yet, admitting so much, we have to feign to ourselves—to imagine hypothetically—an almost infinite number of forms between these two strange creatures and the Bats, that appeared for a time, and then vanished away. Now this free use of the scientific imagination seems to people, generally, to be very much unlike what calm and sober science should be; this is because the minds of many are not in training as to matters of time; they do not use themselves to meditate upon long secular periods. Yet a moment's reflection would show that it is simply impossible, in the nature of the case, that there should be an orderly sequence of phenomena, and no time in which this should take place. This difficulty soon lessens in every well-ordered mind, as soon as the most elementary knowledge of geology is obtained, especially if the echoes of John Milton's song of the creation have been allowed to die out of the inward ear.

The space, right and left, between the Lemurine Aye-Aye, on the one hand, and the unclassifiable Flying Cat, on the other, from the well-ordered and sharply-defined group of existing Bats, is immense. But no one can look at the way in which nature has begun to pull out the Aye-Aye's fingers into bony wires, and to draw out the Flying Cat's skin into a parachute, without imagining further and further degrees of the same processes, which, being continued, would inevitably end in the formation of a Bat. If the palæontologist had not demonstrated these things to us, we should never have conceived of the Pterodactyles, or winged Reptiles, that swarmed in the Secondary epoch; the contemporaries, let me remind you, of the Marsupials of our own region. Then, towards the end of that period, the northern hemisphere abounded in Birds quite unlike our horny-beaked, toothless types. These had jaws well furnished with teeth, some of them had teeth standing against the jaw wall (Pleurodont), whilst others had socketed teeth (Thecodont); both these methods of tooth-fixture still remain in existing Reptiles. Further back a little, in the times of the Upper Oolite, Birds existed with a long tail like that of a Lizard, and with the bones of the hand distinct, as in the Lizard, and not melted together, as in the wing of the modern Birds. The flying Lizard (Pterodactyle) shot out one finger into a long, jointed rod, and to this rod the skin forming the wing was attached, like a sail. We are all familiar with the Bat, and a summer evening's walk would lack one of its charms if no Pipistrelle flitted past our face on his moth-hunting raids. But what an excitement there would have been amongst the astonished palæontologists if the Bats had been all extinct, and just one should have rewarded the labours of a Marsh or a Cope! The Bats are worthy of a whole course of lectures; I can only refer to one or two points in their structure. In the embryo of a small kind of Bat, from North America, no larger than the grub of a Blue-Fly, the curious, expanding hands are rapidly developing, and the folds of the skin are all growing, so that the embryo is quite comparable to the unfolded bud of a plant. Like the Mole at the same stage, this type shows its hereditary characters very early. There is one point in the anatomy of the Bat which puts it on a similar level to that on which we find the Shrew, namely, that there is a considerable rudiment of the coracoid bone attached to the large, keeled "manubrium" (top part) of the breast-bone. This is, as we saw before, the remnant of a structure largely developed in the Monotremes (Duck-bill and Echidna).

The skull of the Bats is intensely ossified and anchylosed, as in the Flying Cat, and as in Birds. There is a chain of small tympanic bones, as in the Crow tribe; besides a cartilaginous bulla tympani (or second inner chamber of the ear-drum), and a bony annulus tympanicus; neither of the two last parts exist in the Bird. But the special adaptations seen in this kind of Mammal for the purposes of flight, are out of the way of any properly ascending survey; they are a sort of side-branch of the life-tree of the Mammalia.

Rodentia (Gnawing Animals).

The Rodents, or Gnawing Mammals, have evidently, like the Bats, arisen from the Insectivorous platform, but they have also undergone a large amount of specialisation; they are still very potent in families, genera, and species. They are, therefore, amongst the most notable of the "winners in life's race," having been able to fit themselves to a great variety of conditions, either by skill or by cunning; by their great fecundity; and also, in many cases, by their small size, so that they have managed to slip through the fingers of their foes. This tendency to dwarfing has been combined with a readiness to pick up any crumbs that may fall from nature's large table. Their teeth are very uniform for so large and varied a group; only the Rabbits and Hares having in the adult state more than two incisors or cutting teeth, even in the upper jaws. They have, apparently, made, as Burns said of his Field-Mouse, only "a sma' request," but this has been, practically, very large. They are a feeble folk, but they are in the habit of going forth in bands and troops, and acting like locusts, whose destructiveness they imitate admirably; that which is peculiar to them is their greediness as well as destructiveness,— "playing the Mouse in the absence of the Cat, to spoil and havoc more than they can eat." Like other mean people if you give them an inch, they take an ell. A good type of the group is the common Grey or Norway Rat; he is one of the skilfullest, and most cunning of the group. But the Beaver is the head or chief of the Order, and his praise is in all zoological treatises. He is at the top: the Guinea-Pig and his huge relative, the Capybara, are at the bottom. This is in conformity with the well-known fact that in the southern world, new or old, neotropical or palæotropical, are found the most archaic or lowest forms.

Thus, in that special territory, the neotropical, we have the Tapir, the Opossum, and the Capybara, and his little, and perhaps still more archaic, relative the Guinea-Pig. This is exactly what I have found with regard to the Bird class: if you want Birds that refuse to fit into well-made zoological systems, that is the place to look for them. I have had my eye upon the Guinea-Pig for many a year, and, working at him from time to time, I have always found something to surprise and stimulate me. The very air of the neotropics must have blown Darwinism into Charles Darwin, when he made that immortal *Voyage of a Naturalist;* there "nature wantons as in her prime,"—there she brings forth out of her treasures things new and old. This feeble-minded neotropical Rodent, the Guinea-Pig, is full of old characters, especially in his skeleton and his skull; and those who study other parts know that he must be made to take the lowest room in his Order. At present, neither in the existing Metatheria, nor even the

Prototheria—the Ornithorhyncus, and his distant relative the Echidna—have I yet found clearer evidence of relationship with the Bird and the Reptile than in this Rodent. Of course, these are not *avian* nor even *reptilian* characters; they are only similar, suggesting a common root-stock for all the three classes. In the palate of the Guinea-Pig I miss the median vomer (ploughshare-bone), the proper vomer of the human anatomy; a hinder and a front pair of vomerine bones exist in this type. In the Mammalia, generally, the pterygoid bones—internal pterygoid plates of Man—are not free from the base of the skull, as in Snakes, Lizards, and Birds, but they cleave closely to it, as in Turtles and Crocodiles. Then, in the Mammalia and the two last-mentioned groups, the skull does not give off outstanding spurs, basi-pterygoids, as in those types which have the pterygoids loose or free. But the Guinea-Pig's skull is like that of Lizard, or Bird, in respect of the basi-pterygoid spurs being well developed, and the pterygoid bones at a distance from the skull. The parasphenoid, or basi-cranial dagger-bone of the Frog, is present as two side pieces, the remnants of the guard; the Flying Cat has only the blade, as I showed in my last Lecture. The small bones of the ear are very large and clumsy in the Guinea-Pig; and many other things in him are archaic, especially the Lacertian condition of the front ribs, in new-born Guinea-Pigs, as I long ago showed, in my Memoir on the "Shoulder-Girdle and Sternum."

The Hare tribe (Leporidæ) are not only non-typical in their teeth, but their front palate, as Mr Howes has shown me, is very different from that of other Mammals. The bulla tympani is not distinct from the annulus, in Rodents, but this part is very large; the Eustachian tube is lined with cartilage derived from the meatus (or cartilaginous outer ear), and outside the bony annulus another imperfect ring is found in some kinds. Outside this the rings of the meatus externus (ear-porch) get less and less distinct, until the large concha (or projecting ear) is reached. The whole structure of the outer ear of a Rodent is similar to the nose-tube in the Hag-Fish, below, among the fishes, and in the probiscidean Rhynchocyon, and the Elephants, above, amongst the high types; the nose-tube is, of course, single in the Fish, but double in the Mammal.

Lemuroidea (Lemurs and Aye-Aye).

The Lemurs are a most puzzling group, no one knows where to put them; they used to be placed with the Primates, because they have four hands (are quadrumanous), but they are now considered to be worth a separate stall. They are marvellously like the Monkeys, but are very inferior to them in every respect, lower in intelligence, and more archaic in type. Therefore, the present Lemurs are not the unchanged descendants of any type from which the Primates sprang; yet the latter, higher, forms may have arisen from old and more generalised Lemuroids. It is just possible that the peculiar isomorphism, or similarity in outer form, of these types with the Monkeys may have deceived us, and have suggested a nearer relationship than really exists. That that modification of the inner digit on both the *manus* and *pes*, hand and foot, which makes these organs capable of grasping, should be seen in both Monkeys and Lemurs, is a strong suggestion of affinity, and yet may not be due to any near relationship. I am inclined to put the two groups nearer together than Professor Flower would; but, as I have only lately begun to work at the Lemurs, I am rather doubtful and cautious. Yet the skull and teeth, as well as the hands and feet, are very similar in both cases, and I feel satisfied that the *Primates* did arise from archaic Lemuroids. The fact that the Lemuroid types found in the Tertiary deposits are undistinguishable from Carnivores, on the one hand, and from the most archaic Herbivora, on the other, shows how difficult classification becomes, when we pass into palæontology. These problems ask for help from embryology; the early conditions of the groups already brought under review, also those of the Lemuroids, as well as the Tapir and the Hyrax, the most generalised types of existing Herbivora, will throw new light upon this question. I hope to do something towards this from year to year, but many hands will be wanted to make light work of this question. One kind of Lemur, which I have studied in a sub-adult condition (*Cheirogaleus Smithii*) from Madagascar, is much smaller than a Norway Rat, and yet its skull has a very *human* look. Another, a larger kind (*Lepilemur*), pleases me vastly: —it has a very curious dentition in front, like that of a Ruminant—Sheep or Cow.

In this kind the upper incisor teeth are nearly all suppressed, and the lower are all like chisels; this dentition is quite similar to that of our familiar even-toed cattle; and, if found in some generalised type of ancient herbivorous Lemurs, would assuredly be set down as a prolepsis, or anticipation, of those noble recent beasts. Madagascar is the great home of the Lemuroids, although they are found on the mainland of Africa and Asia, and on certain islands contiguous to these continents. The most startling form of the Lemuroids is a native of Madagascar; this is the Aye-Aye, *Cheiromys*. In this type, in the adult, the dentition is very similar to that of the Squirrel, whilst the fingers are elongated into long, pointed structures, that seem to be the wing of a Bat, *in process of making*. Some say that the Aye-Aye pokes large grubs out of deep holes in trees with these curious fingers; I have seen him, through Mr Bartlett's kindness, dip his long knuckles into a sweetened, half-liquid diet, and then, as though he were a cook, draw the convex surface of the digits across his partly opened mouth; thus he would appear to live by licking his fingers.

Carnivora.

The types now to be mentioned are the nobler Eutheria; these have gone on improving, through secular periods, not easily measurable, but certainly very great. The Carnivora have been well worked at by Professor Flower; to his various and valuable papers and works, I must refer you. But that which is most curious in the skull of the Carnivore, the special "bulla tympani," comes into my particular department, and I may now explain its meaning. I have mentioned that the meatus externus, or outer passage of the ear, the "porch of the ear," in Shakespeare's language, is formed of a superficial cartilage, which reaches through the Eustachian tube to the throat. This is segmented into imperfect rings, especially in its inner part; the innermost ring but one is tranformed into bone very early, before the cartilage gets solid. This becomes the annulus tympanicus, or tympanic bone. The innermost annulus becomes a truly cartilaginous tract in the Carnivora; it then ossifies, independently, and after that coalesces with the already ossified bony ring, next outside it.

These Carnivores are very tempting to the biologist, but they do not lead towards the human race. Yet they are cater-cousins to savage Man, to the natural Man, pure and simple, in reality neither pure nor simple.

Hyracoids and Proboscidea.

An embryo of the Elephant is still a desideratum to me, in my special work, and so is that of the Hyrax—the little Syro-African Coney, not *Cuniculus* or Rabbit. My esteemed friend, the late Professor Rolleston, sent me the skull of a new-born Hyrax, and from that I learn that no large amount of cutting and contriving will be necessary to make this type of skull fit in with that of other low and generalised forms. With this skull I shall in due time compare that of the Chevrotains or Tragulidæ, the older types of Ruminants.

Cetacea and Sirenia.

As to the Cetacea and Sirenia, there must be much slow and cautious work and thinking before we shall be able to relegate them to their proper place in the class.

Palæontology is as unready here as embryology. That these forms have undergone great changes during long periods of time, there can be no doubt; they have many of them lost their teeth; all have lost their hind limbs; their fore-face has grown into a huge beak, and their bulk has become simply monstrous. The Sirenia —Manatee and Dugong—are as great a mystery as the Cetacea. They are almost extinct; the largest of them, the *Rhytina*, of the seas north of Asia, has only lately fallen from the ranks of life.

Quadrumana.

I shall leave the Quadrumana in their awkward and disagreeable contiguity to the ruling race. Nothing is so unpleasant as to be pestered with poor relations, especially if they be foul in feature and rude in manners. Surely these old people of the land have been left among us—nature not driving them all out at once from the human Canaan—that they might be as thorns in our sides, and as goads and snares, that we may be pricked on to high courage, and strong resolves, in our moral work of extinguishing all that is Ape-like in our nature. For my own part, when, as a biologist, I walk abroad for pleasure, it is not into the "back-slums" where these Mammalian Satyrs live ; and if it be my lot to undergo metempsychosis, I shall pray to be transformed, rather into a Barnacle, than into an Ape,

"With forehead villanous low."

Suidæ and Hippopotamidæ.

The *Pig* is not a lonely creature; though much generalised, he comes nearer the special form of the Ruminant than any other outside that group; his only rival is that huge, unwieldy beast, the Hippopotamus—probably Job's "Behemoth." The Pigs are numerous, but the Hippopotamus is one of a sub-extinct family. These forms have lost the thumb and great toe, but have retained the normal distinctness of the metacarpals and metatarsals. The teeth have not, in either case, run up to the perfection of pattern seen in the Ruminants and Solipeds. Some years since, I published, in detail, an account of the development of the skull in the Pig; that paper still holds its place in my affections, if not in those of other anatomists. I hope soon to add a similar memoir on the skull of the Ruminants, and another on the skull of the Horse.

Ruminants.

Correlated with the specialisations that have taken place in the Ruminants, throughout the body generally, there are some very curious things to be noticed in their skulls; I shall not, however, trouble you with details. The skull of the Chevrotains, especially that of the Hyomoschus, the African kind, which has the two middle metacarpals distinct, will be profitably studied as that of a low and almost extinct type of the Ruminants—looking Pig-ward. The general anatomy of these herbivorous forms is not my work. Professor Flower is *facile princeps* there; as also with the Whales, &c. Our distinguished predecessor (Professor Flower's and my own) in this Chair, Professor Huxley, has done most to make the *pedigree* of the Horse plain to us; and to his paper on this subject I must refer you. I will, however, read one pertinent and powerful passage out of one of his latest papers, and thus conclude this rambling Lecture.

Professor Huxley on the Evolution of the Horse type.

" . . . . . Since the commencement of the Eocene epoch, the animals which constitute the family of the Equidæ have undergone processes of modification of three kinds—

" 1. There has been an excess of development of some parts in relation to others.

" 2. Certain parts have undergone complete or partial suppression.

" 3. Certain parts, which were originally distinct, have coalesced.

" Employing the term 'law' simply in the sense of a general statement of facts ascertained by observation, I shall speak of these three processes by which the *Eohippus* form has passed into *Equus* as the expression of a threefold law of evolution. . . . ."—" On the Arrangement of the Mammalia," *Proc. Zool. Soc.*, 1880, p. 649.

LECTURE IX.

CONCLUSION.

THE hardest part of my task has yet to be done. There is one thing, however, comfortable to you and to me in the matter; we may now "take courage, for we see land." Once through these stormy straits, we shall come to a haven of rest.

We can then meditate upon our mysterious origin at leisure, making ourselves certain of this one thing, that we are what we are; no theorising as to our origin can upset that fact.

You very much mistake me if you think that I make a boast of being an organism—a compound of flesh and blood, bones and nerves. If the manifestation of what we generally call "spirit" has hitherto taken place through the medium of matter in some organised form, it may still be hoped, that as it appears to be the most homogeneous thing in us, it may survive the breaking up of the organised shell which now imprisons it.

A frightful attack of spasms, passing into tetanus, falls suddenly upon the "pupa" of the Dragon-Fly. In the agony of his last moult (ecdysis), he little conceives of the glory that is to be revealed in and to him.

What must seem to him to be the very bitterness of death is, in reality, the passage from the shadow of death into a newness of life, as delightful as it must have been unexpected.

A word to the wise! The great apostle of metamorphosis, our beloved Darwin, never doubted, and never dreamed of doubting, the future possibilities of our mysterious being, when he showed us that, in our initiation into life, we were of the earth, earthy, and like unto the beasts that perish. "Sit anima mea cum philosophis!" May my soul always walk with his soul, and with the souls of such calm, gentle, noble, and expectant natures as his!

Anyhow, in the very delirium of life's fever we have had visions of marvellous beauty; beauty developed upon this little, dark terrestrial ball. The possibilities of the universe are practically infinite, and we can afford to wait until the time of our "ecdysis" comes.

But some one will say, should not all this speculation and controversy about the evolution of organic types—with MAN at the top—be avoided, as likely to fill us with troubled dreams?

We cannot avoid it; it is in the air. Once lift up Man above the Tiger and the Ape, and he will begin to speculate. I would not exchange Darwin's *Origin of Species* for all that the "schoolmen" ever wrote; and write indeed they did—the world groans under the books they have written. I should be sorry, indeed, to be obliged to read "such skimble-scamble stuff, as puts me from my faith."

We are still as much in the dark as ever as to the nature of the soul, though we do know more about the body than those same "schoolmen" did. They knew as much as we, and we as much as they, about the *size, form, weight,* and *colour* of the *soul*; but we alone can trace the body through its mysterious development to its final decay.

Not only could any number of "souls" be blown through the eye of a needle,—but the germinant tenement of the soul, in its first inception, in other words the tiny germ, that is the MAN *in potentia*, is smaller than anything we can imagine as a visible object. But the growth and expansion of this mysterious life-point has, however, been traced by trained eyes, assisted by all the means and appliances of modern embryological science.

I must ask your attention whilst I mention some of the stages of this personal or individual evolution. I must indeed repeat several things in this that have been asserted in other Lectures delivered here —in this, and in former years. Still, since my learned hearers humble themselves, and sit as though they were pupils, they must expect to have "line upon line, precept upon precept; here a little, and there a little, time after time."

Is there anything in the human organism of which it can be said, "See, this is new?" No; "it hath already been in the old time that was before us."

We modern English, I think, ought to be put somewhere between the old Greeks and the ancient Jews in any classification that should be made of the race, according to its mental characteristics. The Jews required a "sign," the Greeks sought after wisdom; we clamorously ask for the former, whilst groping after the latter. It is just possible that with these mental modifications the finest and happiest breed of Men may be brought into being after the long ages of human evolution, and that the latest birth of science will be that which will satisfy and satiate the longings of this mixed kind of Man.

I see in every flower, in every insect, that which absolutely transcends, not only the highest human skill, but our utmost efforts of thought and flights of imagination. These insects are highly complex organisms; but lower organisms are really as wonderful. I spent some of the best years of my life in conjunction with scientific friends in the study of the "Protozoa," or simple primary animal forms, and we found that the apparently simple flesh-stuff or protoplasm of which these creatures are formed could do such wonders of quasi-architectural growth, as made all our human cleverness but as the cleverness of a speaking Fish or a wise Pig.

Now the Protozoa end where we begin; we are

"Metazoa" with a vengeance; we become differentiated, and get organs of various sorts for diverse uses, physiological and otherwise; and all these various organs, in their manifold physiological action, are correlated, so as to work together for the good of so complex an individual as a living Man. It is indeed a good and pleasant thing to see how all these brother-organs work together in unity; how they are all set in double ranks, so to speak, and one part is made to serve the good of another. Thus the whole frame, made of parts compacted together, fitly united by joints and bands, and having living fountains of nourishment ministering to every part, groweth into a temple fit for an indwelling DEITY.

The earth is one kind of "organism," the body another; they are comparable; and the pulses of the sea remind one of the waves of the heart. In each case the contemplation of these actions is full of charm to the intelligent mind. Those whose minds are attuned to this harmony feel it everywhere; to these, the stars in their motion sing, and the river in its course. But "the current that with gentle murmur glides, giving a gentle kiss to every sedge he overtaketh in his pilgrimage," is surely not fuller of harmony and music than the pulsating artery that sends the colour "to the lovely lady's cheek." Yet this whole body was, a few years ago, a speck of protoplasm, practically a "monad," or a protozoon. Here, from this standpoint, we can contemplate the two beginnings—1st, the great pro-cosmic beginning, when the foundation was formed of the earth nascent, when the sea was born, and had the soft clouds for its swaddling-bands; and 2nd, the lesser protoplasmic beginning, the first budding of the tiny germ of the individual human being. This latter is a historical repetition of the former,—it is the beginning of but one among many millions of kinds that developed from the first great protoplasmic root of organisation; but that one explains all the others—gives the reason of existence for all the others.

You may put the whole organic world into two categories if you like, Man on one hand, and all other living forms, vegetable and animal, on the other; thus you have MAN and *not Man*.

Now nature seems to have loved the former more than the latter; by election she has made him heir of all eternity; she has taken, so to speak, one tribe of creatures from the midst of the other tribes, and set them on high above them all, so that they have become a "peculiar people," bringing all the rest into subjection to themselves, either killing them outright, or, if permitting them to live, making them hewers of wood and drawers of water. But this killing and culling goes on with the human family, the strong and the wise destroying the weak and the foolish; and looked at from one standpoint, this race seems to be only an improvement upon the unreasoning races by its superior skill in the art of destruction.

Is this, then, "the be all, and the end all" of organic life upon the planet; or will the races that become the noblest turn round, go clean contrary to their progenitors, the blood of their fellow-men becoming precious in their sight? If this should take place, and the finer kinds of Men, rich in strength and life, and all the means and appliances of life, should spare the poor and needy, and save the souls of the needy, then this counter-evolution will be a revolution indeed.

History throws some light upon this dark problem, our Celtic, Saxon, and Danish forefathers, the Wolves and Bears and Eagles of the race in these northern parts, —Men who surnamed themselves by the names of those cruel creatures,—had, after a time, the sentiment of mercy ingrafted into their wild minds; some at least of their descendants show us how rich are the results.

I have broken my scientific tether, and have got, somehow, into the region of ethics. I am sure that you will sympathise with me, for I was thinking of what Man would be likely to come to, ultimately, on mere necessitarian theories of evolution. If we all were mere biologists, pure and simple, then any lament over the destruction of low types, or any sentimental care for the poor and the afflicted, would be absurdly out of place in our observations and deductions. But in most of us there is a strange mixture of the natural and the mystical—of the child's joy in finding facts and drawing inferences, and the old man's sorrowful reflection that his days upon

o

earth are but as a shadow. Yet surely this fog of mixed thought and sentiment ought not to sadden and to paralyse us; let us hear the wise man once more :—" Then I returned and saw that wisdom recedeth from folly, as far as light recedeth from darkness; for the wise man's eyes keep watch in his head, but the fool roundeth about in darkness."

Now, all of us whose tendency is toward the sentimental and the mystical—who wish to travel along the path that no fowl knoweth, and that even the Eagle's eye hath not seen—are in danger of undervaluing "the light of nature."

For the third time within three hundred years the teachers of the mystical are reconsidering their views as to positive science. Astronomy, Geology, Biology— especially the last—have asserted their claims, have maintained their rights, and have conquered their bitterest opponents. The earth is not flat, nor fixed, nor the centre of the universe. The stony foundations of the earth—the ancient mountains, the pleasant hills—were not juggled into existence in a moment of time. Nor did the living creatures come up out of the earth— shivering, like the ghost of Samuel, and shaking the mould off their backs, all literally at the word of command.

That the grand old "song of creation" should have been so misconceived, is one of the greatest humiliations to the intelligent mind; yet the mistake arose through our forgetfulness of the fact, that in a drama the element of time is often annihilated for dramatic purposes. There is Darwinism in this also; the cold north, which has hardened our bodies and made us the envy of all nations, in iron energy and cool courage, has also chilled our imagination and stiffened our mental frame. Yet, cold and unimaginative as we are, we shall never cease to be mystics, for we are floundering in a sea of mystery : but there are subjects that our minds can grasp—things that our hands find to do ; let us do them with our might.

Even the ancient heathens, says Bacon, believed that the highest link of creation was fastened to Jupiter's chair; are not our faculties given to us that we may trace the golden links one after the other, until we, mentally, reach that throne?

Are our faculties trustworthy? I think they are. Is there anything in creation more sacred than the human mind,—any *conditioned Being* that comes nearer to the UNCONDITIONED BEING? May not human beings, by a process of development, go on improving, so as to become what the Eastern folk would call "sons of light?" These are questions to be asked; let us answer them by acting.

Whilst the world remains, there will be mystics and naturalists; there will be dreamy metaphysicians, and active, bustling men of science. Such opposite types will always be a trial to each other; if they are pulling at the same plough, we see at once that they are unequally yoked together. Sometimes we get a "cross" between the two, but the two halves of such a nature seldom amalgamate well; yet there are some many-sided minds that do wonders for us in harmonising even these opposites.

There is no room for despair. Our language has become enriched, since some of us were young, with many of the most valuable words that ever came in to help the development of thought. Our knowledge has proceeded beyond the knowledge of our progenitors, stretching, not merely to the utmost bounds of the sea-girt hills, but to the bounds of the Cosmos, so to speak, which has no bounds; and diving into the depths of the warm life of living Being. This branch of human knowledge has, I suppose, grown more during this nineteenth century than in all the times before; back to Aristotle, back if you like to Moses, with his clean and his unclean beasts. This is a new kingdom of science, this embryology, but you have to enter it through a strait gate and a narrow way; there is no royal road to it. For here, unless a man stoops his head, he will bruise it; unless he enter silently he will learn nothing. It is not nature revealed in a strong wind, or an earthquake, or a fire; it is the still small voice of the growing cells he must train his ears to hear.

"The spectacle afforded by the wonderful energies prisoned within the compass of the microscopic hair of a plant, which we commonly regard as a merely passive organism, is not easily forgotten by one who has watched its display, continued hour after hour, without pause or sign of weakening. The possible complexity of many other organic forms, seemingly as simple as the protoplasm of the nettle, dawns upon one; and the comparison of such a protoplasm to a body with an internal circulation, which has been put forward by an eminent physiologist, loses much of its startling character.

"Currents similar to those of the hairs of the nettle, have been observed in a great multitude of very different plants, and weighty authorities have suggested that they probably occur, in more or less perfection, in all young vegetable cells. If such be the case, the wonderful noonday silence of a tropical forest is, after all, due only to the dulness of our hearing, and could our ears catch the murmur of these tiny maelstroms, as they whirl in the innumerable myriads of living cells which constitute each tree, we should be stunned, as with the roar of a great city."[1]

"I was curiously wrought," says the psalmist,— wrought as with the fine needle of some skilful Eastern maid, who makes "apples of gold in pictures of silver." Any one who will watch the wonderful process of cell-growth, and follow it through all its mysterious silent passages and labyrinths, as seeing things that are in-

[1] Huxley, *On the Material Basis of Life*.

visible, must have the skill and patience of "the lady of Shallot." Moreover, all this patient skill is of no avail, if there be no interpreting, and even prophesying power in the mind; an embryologist should be a Seer.

Now, what will he learn? First of all, he will learn humility—that is the first lesson: "I will say to the worm, Thou art my mother and my sister;" this, one of the lowlier of the living tribes, is half-way up between Man and his protozoic ancestors. The labour of the embryologist may be likened to that of a scholar who finds his work at one time in ransacking a huge Bibliotheca, and at another in scanning the pages of a delightful abstract of a huge folio, where the gist and essence of the subject is put as in a nutshell.

The whole organic kingdom is the Bibliotheca; Man is the consummate abstract; he "seals up the sum," being, in his perfection, full of beauty, if not always "full of wisdom." Man may be said to be an excellent manual on the teleology—the final purpose—of creation.

If the body of man be considered as the vehicle of his mind, a ship which carries him, but of which he is the governor or pilot; then of that ship or ark it may truly be said, a great while was it in building,—a far longer time than it took to construct the lesser and meaner kinds of craft—mere boats as compared with the nobler vessel. Such a vessel is the last result of the working of the morphological force, and may be compared to that ship which carried Cæsar and his fortunes; for MAN, in the widest sense, may be likened to that notable personage who came, saw, and overcame. The lust of dominion is in the blood of this born ruler; and the bold poet of the Cosmos—"that chosen shepherd who sang how the earth rose out of Chaos," makes the Divine Interlocutor say—"Have thou dominion over the fish of the sea, and over the fowl of the air, and over every living thing that moveth upon the earth."

He, indeed, was sent here naked, and with all instruments of defence and offence taken from him, or so filed down and lessened as to be next to useless, considering the folk he had to fight against and overcome. But in the folded volume of his huge brain, all things were written that were necessary for him in this dread emprise; he has conquered, because he could learn how to conquer. "Who made you, my child?" asks the solemn, puritanical Miss Phely, of the little negro girl Topsy— two of Mrs Beecher Stowe's charming creations—"Specs I growed" is the quickly given answer. That answer contains as much "good theology" and sound doctrine as the more usual answer to this part of the Puritan catechism. "Whilst this muddy vesture of decay doth grossly close us in," we shall get no nearer to the solution of problems of this sort, as seen from the metaphysical side.

But on the organic or physical side, we are always gaining fresh territories; here, strong and brave minds are ever going on, conquering and to conquer. And the more mystical and inward sort of thinkers are becoming more and more indebted to the men who like to have things plain and above board, men who stick to facts and phenomena. I much question whether there is a single modern work of any worth on any subject whatever, on mind or matter, that is not the better for what Charles Darwin and his helpers and interpreters have done. As there were kings before Agamemnon, so there were biologists before Charles Darwin. A century ago, his own grandfather, Erasmus Darwin, had begun to break the yoke and burst the bonds that bent and bound the minds of men. To Goethe, however, must be attributed the honour of being the father of all those who treat of Morphology and Development.

When the eyes of the prophet's servant were opened, he saw no longer barren rocks with mist resting upon them, but the whole mountain was full of chariots of fire, and horses of fire. The vestments and ritual of nature may take up all the attention, and use up all the energies of her votaries; these superficial observers fail, however, to find the real religion of nature—the beautiful but awful omnipresence which every flower and every insect reveals. The phenomena of nature are all mere fading pageants, and the really cultivated mind only finds lasting satisfaction in meditation upon the recognisable forces that underlie all sensible appearances.

This, however, is what the old philosophers called "dry light;" and, as Bacon remarks, it is not comfortable to most minds. The deeper things of nature are a sort of manna, but the souls of some people become dried up if you give them merely this celestial kind of diet; so that they murmur, and say, "We remember the fish which we did eat in Egypt, freely; the cucumbers, and the melons, and the leeks, and the onions, and the garlic."

And yet this ignorance of nature is set up as a dead wall against all progress of thought; for these people are "most ignorant of what they're most assured," certain that they know all about their "glassy essence;" and, although as blind as moles, they are the enemies of all who have had their eyes opened, to whom the mountain is no longer misty and dark, but flaming with light.

"Ne sutor supra crepidam"—do not trust the cobbler in things outside his calling—is a proverb that cuts both ways. The biologist may surely be allowed to know things that relate to his own calling: the man who never dreams of life, and the science of life, should be careful how he contradicts its experts. On the other hand, bigotry is not confined to one class of controversialists; some very bitter things have been said against faith by men whose culture and science ought to have taught them better. We have a right to look for nothing but "sweetness and light" from the apostles and prophets of this new dispensation.

When the dust of controversy shall have subsided, when those who have to receive new ideas as if by a surgical operation begin to feel the stirrings of these new conceptions thus let into them—the "new leaven" of nobler thoughts about nature, and of the great First Cause of nature—then all who can think will find that they are colonising a "new Atlantis."

The "old song" of the creation puts it thus,—"Evening was—morning was,—day one."

Thus the shadows of evening come first, and the rosy light of dawn afterwards. Now, in science, even in biological science, the "morning is spread upon the mountains," and soaring birds are singing at heaven's gate; so that the drowsiest folk are beginning to stir themselves, "ere well awake."

ADDENDUM TO LECTURE IX.

Lord Bacon, in his *Advancement of Learning*, is very severe upon the metaphysicians of the Middle Ages. I must be allowed to give a few of his pithy sentences; take the following :—

"Surely, like as many substances in nature which are solid do putrefy and corrupt into worms; so it is the property of good and sound knowledge to putrefy and dissolve into a number of subtle, idle, unwholesome, and, as I may term them, vermiculate questions, which have indeed a kind of quickness and life of spirit, but no soundness of matter or goodness of quality. This kind of degenerate learning did chiefly reign amongst the schoolmen, who, having sharp and strong wits, and abundance of leisure, and small variety of reading, but their wits being shut up in the cells of a few authors (chiefly Aristotle their dictator), as their persons were shut up in the cells of monasteries and colleges, and knowing little history, either of nature or time, did, out of no great quantity of matter and infinite agitation of wit, spin out unto us those laborious webs of learning which are extant in their books. For the wit and mind of man, if it work upon matter, which is the contemplation of the creatures of God, worketh according to the stuff, and is limited thereby; but if it work upon itself, as the spider worketh his web, then it is endless, and brings forth indeed cobwebs of learning, admirable for the fulness of thread and work, but of no substance or profit."

As to "second causes," and the Divinity who is enthroned above and beyond the Cosmos, his remarks are so pertinent and so full of eloquence and wisdom, that I cannot refrain from further citation. He had been quoting "One of Plato's School," as to the obscuration of divine things by objects of "the sense," and then he goes on to say—
"And hence it is true that it hath proceeded, that divers great learned men have been heretical, whilst they have sought to fly up to the secrets of the Deity by the waxen wings of the senses. And so for the conceit that too much knowledge should incline a man to atheism, and that the ignorance of second causes should make a more devout dependence upon God, which is the first cause; first, it is good to ask the question which Job asked of his friends, Will you

lie for God, as one man will do for another, to gratify him? For certain it is that God worketh nothing in nature but by second causes; and if they would have it otherwise believed, it is mere imposture, as it were in favour towards God; and nothing else but to offer to the Author of Truth the unclean sacrifice of a lie. But farther, it is an assured truth, and a conclusion of experience, that a little of superficial knowledge of philosophy may incline the mind of man to atheism, but a farther proceeding therein doth bring the mind back again to religion: for in the entrance of philosophy, when the second causes, which are next unto the senses, do offer themselves to the mind of man, if it dwell and stay there, it may induce some oblivion of the highest cause; but when a man passeth on farther, and seeth the dependence of causes, and the works of Providence; then, according to the allegory of the poets, he will easily believe that the highest link of nature's chain must needs be tied to the foot of Jupiter's chair. To conclude, therefore, let no man, upon a weak conceit of sobriety or an ill-applied moderation, think or maintain that a man can search too far, or be too well studied in the book of God's word, or in the book of God's works; divinity or philosophy; but rather let men endeavour an endless progress or proficience in both; only let men beware that they apply both to charity, and not to swelling; to use, and not to ostentation; and again, that they do not unwisely mingle or confound these learnings together."

The most capable of those whose studies lie in the domains of philosophy and theology are quite willing that the biologist should carry his own proper researches as far as his faculties and opportunities will permit. Dr Westcott's views on this matter are well worthy of attention:—

"There can be no antagonism," he says, "between Theology and Science as they are commonly contrasted. So far as these keep within their proper limits, they move in distinct regions. Their respective paths lie in parallel, and therefore in unintersecting, planes. Theology deals with the origin and destiny of things: science with things as they are according to human observation of them. Theology claims to connect this world with the world to come: Science is of this world only. Theology is confessedly partial, provisional, analogical in its expression of truth: Science—that is, human science—can be complete, final, and absolute in its enunciation of the laws of phenomena. Theology accepts without the least reserve the conclusions of Science as such: it only rejects the claim of Science to contain within itself every spring of knowledge and every domain of thought." —*The Gospel of the Resurrection*, 3rd edit., p. 48. London, Macmillan & Co., 1874.

Study of final causes does only, as Bacon says, "slug and stay the ship." Of course, the watch did not make itself, but its works were not more certainly intended to mark the lapse of time, than a Beaver's teeth to cut down the willows of the brook. But we get a curious mixture of folly and irreverence when the poor human watchmaker is compared to the Power that lies behind all organic machinery.

If we compare the structure of the organs of any particular group of animals, we see that each minor variation has reference to some peculiarity in the habits of the creatures so varied. But the Darwinian believes that these modifications took place, in time, through the influence of the surroundings upon a delicately balanced organism, ever sensitively alive to the influences that play upon it—ever ready to respond to the wants and instincts of the creature.

I have just spoken of the Beaver; he is extinct in this country now; but some small relatives of his are at hand for illustration. The Rats and Mice of this country are divisible into two groups, namely, those with comparatively simple grinding teeth, not unlike our own in miniature, and those whose grinders are very much like miniatures of the grinders of the African Elephant. The Grey or Norway Rat, the almost extinct Black Rat, the Domestic Mouse, Wood Mouse, and Harvest Mouse (all members of the genus *Mus*),—these all have those simpler tuberculated grinders, capped with enamel, and having short roots or fangs in the adult. But in the Water Rat (*Arvicola amphibius*) and in the short-tailed Field Mouse (*A. agrestis*) the grinders, like the incisor teeth of both groups, have permanent pulps, and their enamel is folded in among the ivory (dentine), and these two substances, thus enfolded, are surrounded by a sort of bark of bone called cement—a substance that is formed in the roots of the simpler teeth in the other group, but in smaller quantity. When worn, these grinders have a W-shaped pattern, and these small millstones keep the roughness of their surface by the faster or slower wear of the three kinds of stony substance of which they are composed. The cement wears fastest, the ivory next; the enamel is the hardest, and wears very slowly; thus elegant ridges of this tooth substance are always seen on the worn surface of the tooth. There the "final purpose" is not far to seek; the question is how this modification came about;—a modification correlated with numberless other differences of structure that run through the whole animal, and are in harmony with one another. Those with the permanently-growing compound grinders (the genus *Arvicola*) are almost pure vegetable feeders, like the Beaver; those with simpler grinders, the members of the genus *Mus*, are omnivorous with a vengeance. The Norway Rat is called *Mus decumanus*, or tithe-farmer; he does not *pay* tithes of all he possesses, but he *takes* tithes of all you possess, and he is as unscrupulous a rogue as any of those men with itching palms whose title he has had conferred upon him. Now, if a Darwinian were asked whether the difference between the teeth of these two types of Rodents was a case worthy the interposition of the "Deus ex machinâ," he would be ready to laugh at the simple question. He cannot certainly bring out of their graves the whole kin of the Arvicolan forefathers to prove that very long ago they had all a simpler kind of grinders, like the grinders of the existing genus *Mus*; but if you will accompany him to any Natural History Museum, he will show you every intermediate condition between the two types of grinders, and make it plain to you as to the easiness of the transition. Gentle modifications of this kind; very small variations accumulating during long secular periods, in every case in harmony with the habits of the creature, slowly, or even suddenly, changing with changing conditions of temperature, food, fresh needs for safety, and the like;—all these things have to be taken into account.

It is reassuring to find that Goethe held very similar views to those put forward in these lectures.

Eckermann, that German Boswell, has preserved for us the following remarks on the doctrine of "purpose," which will bear quoting here:—

"Goethe has been speaking of the book of a young natural philosopher, which he could not help praising, on account of the clearness of his descriptions, while he pardoned him for his teleological tendency.

"'It is natural to man,' said Goethe, 'to regard *himself* as the final cause of creation, and to consider all other things merely in relation to himself, so far as they are of use to him. . . . He cannot conceive that even the smallest herb was not made for him; and if he had not yet ascertained its utility, he believes that he may discover it in future. Then, too, as man thinks in general, so does he always think in particular, and he does not fail to transfer his ordinary views from life into science, and to ask the use and purpose of every single part of our organic being.

"'This may do for a time, and he may get on so for a time in science; but he will soon come to phenomena where this small view will not be sufficient, and where, *if he does not take a higher stand*, he will soon be involved in mere contradictions.

"'The utility-teachers say that oxen have horns to defend themselves with; but I ask, why is the sheep without any, and when it has them, why are they twisted about the ears so as to answer no purpose at all?

"'If, on the other hand, I say the ox defends himself with his horns because he has them, it is quite a different matter.

"'The question as to the purpose—the question, *Wherefore?* is completely unscientific. But we get on further with the question *How?* For if I ask *how* has the ox horns, I am led to study his organisation, and learn at the same time why the lion has no horns, and cannot have any.

"'Thus, man has in his skull two hollows which are never filled up. The question *wherefore* could not take us far in this case, but the question *how* informs me that these hollows are remains of the animal skull, which are found on a larger scale in inferior organisations, and are not quite obliterated in man, with all his eminence.

"'The teachers of utility would think that they lost their God, if they did not worship Him who gave the ox horns to defend itself. But I hope I may be allowed to worship Him, who, in the abundance of His creation, was great enough, after making a thousand kinds of plants, to make one more, in which all the rest should be comprised; and after a thousand kinds of animals, *a being which comprises them all—Man.*

"'Let people serve Him who gives to the beast his fodder, and to man meat and drink as much as he can enjoy. But *I* worship Him who has infused into the world such a power of production that, when only the millionth part of it comes out into life, the world swarms with creatures to such a degree that war, pestilence, fire, and water cannot prevail against them. That is *my* God.'"—Eckermann, *Conversations with Goethe*, translated by Oxenford (year 1831, vol. ii. p. 347).

General remarks on the unity of structure of the Mammalia as a group, Man included, are apt to evaporate from the memory. I will give an instance of the marvellous conformity of structure between Man and one of the beasts, that came under my own notice, and strongly impressed me, many years ago.

In the year 1845 or 1846, a large Seal (*Phoca vitulina*) lost its way by entering the mouth of the Welland, and was caught at Stamford, in Lincolnshire. It was kept as a show for a while, and the descendants of the "holy-day fools" of Shakespere's time paid their monies to see the monster.[1] This captive soon died for want of a due supply of fish, and was purchased by me for dissection. The skeleton of it is still to be seen in the Hunterian Museum. This gentle creature, the Seal, is, as every one knows, very fond of music, and looks as though he were one of our own kind, mysteriously bewitched into a fishy shape, as though he had "suffered a sea-change" wrought upon him by some demon of the deep. His eyes are so soft and tearful, and have such a sorrowful, longing look, that you seem to hear him ask you to break the spell that keeps him bound in his watery prison, his hands and feet tied up, so as to be of no use except as swimming flappers.

This bewitchment theory of the cause of the poor Moon-calf's shape, and enforced watery life, is borne out by his structure in the most remarkable manner; many worse and more absurd scientific theories than that have been hatched in human brains. Now this is what I learned of the conformity of the structure of this creature with that of Man, namely, that it would be possible for an accomplished anatomist to write an accurate account, using somewhat general terms, but naming every muscle, bone, nerve, artery and vein—every senseorgan, and every one of the soft viscera within its body—and that description might be made to serve both for the Seal and for Man. More than this—far more—the same supposed biologist

[1] *Tempest*, Act ii., Scene 2nd.

might take up the embryology of the Seal, tracing the confluence of the two primary germ-points, their mutual engrafting, the growth of the foundations of the embryo (the " blastoderm "), then the differentiation of the various tissues and organs—and this second piece of descriptive anatomy might serve equally well both for the Seal and for Man. Then if the physiologist took up the subject, the functions of every part would be found to correspond, and the physiology of a Seal would be seen to be essentially the same as the physiology of our own more favoured type.

I do not ask the reader to go through all the details and experiments for himself; but he might waste an hour in a less pleasing and profitable manner, than by comparing the Seal's skeleton with that of his own species, in that model Museum in Lincoln's Inn Fields. Then he would see in the skull, the spine, the chest, and the limbs, part for part, joint for joint, bone for bone, the same structure, but just gently altered, for somewhat different functions; altered as if by the hand—not of a demon of the deep, but by a kindly fairy—so little difference is there in the details of the skeletons of two creatures, so diverse as a Man and a Seal.

Every organ, every part in the adult creature, Man or Seal, or indeed of any complex organism whatever, has its *time-marks* upon it, within it, and throughout it. In these northern climes, where there is one growing and one resting season every year, an ordinary tree that has been growing for 200 years will show, when cut across, 200 rings of wood. But a *long bone* from our own body, say the thigh-bone (femur), if sawn across, would show similar growth-rings to those seen in the tree when sawn across. Of course, as we are not fixed trees, but active men, not standing as if we could not find our hands, but supplying ourselves with food constantly, our growth-rings do not depend upon the seasons, though they are the signs of the lapse of time. Now, between the growth of the tree and that of the thigh-bone, there is a remarkable similarity and an equally remarkable difference. The soft tissues of the embryo of a tree, the oak, for instance, which grow into the axis or rudimentary stem, gradually become differentiated into various kinds of cells, which form the pith, medullary rings, large ducts, woody fibres, and the inner and outer layers of the bark. Year by year an additional ring, or tube of wood, is formed outside the last year's growth, and inside

P

the bark (that is just where the sap abounds most—the moisture which we see when we peal a green withy, exposing the wood). If this concentrically increased stem becomes hollow after a time, that hollowness is due to some injury, such as the breaking off of a branch, so that the wet from without drips into the wound, and thus rots the inner wood. This inner wood, the "heart-wood" as it is called, is dark in colour, being filled with a dark deposit; it has ceased its vital functions, and is only kept sound by being hermetically sealed by the newer sap wood, or white part, and the bark outside this.

Now the thigh-bone of a man is as hollow as an old tree, save that it is filled with marrow, the oily substance that fills the cavity of every long-bone. The wall of the bone is in concentric rings or tubes like the tree, and the part that is gone was in concentric rings also. How did this come about; where is the small bone of the child which did lie where there is only a useless sort of padding now —nothing but marrow? We are here in the presence of one of the most marked distinctions between the animal and the plant, the Man and the Oak tree. Plants have not the power to dissolve old tissues, using up the nutritious parts of a solution so made, and casting the rest off,—excreting the effete and useless matter; but animals, especially the higher kinds, can do this—are always doing it; thus every tissue in the body is busy as a hive of bees. For a time, if a child or youth be well fed, the deposit of new substance exceeds the removal of the old, and the body increases in size throughout. Leaving out of consideration, for the time, the rest of the body, I may remark that the organs of support undergo a most remarkable series of changes from the time when they are differentiated in the early embryo to the time when the adult condition is attained.

At first all is mere protoplasm; of the embryo, it may be said, parts it has none, distinguishable in member, joint, or limb; each seems either; all appears to be uniform. After a time granules appear, and abound; these become cells, or little living pellets of protoplasm, instead of the undistinguishable jelly; and, a little further on, these are marshalled into ranks, and divided into groups, so that we seem to have a silent sort of building going on, with infinitesimally small bricks or stones. These, however, are not shaped outside and then put together; they model themselves within, where they are; there is no sound of hammer or of axe in the uprise of this living building. The very term "building," which we can scarcely avoid using, conveys unconsciously a wrong conception of the process; a better idea of the process is gained by saying, that we are nature's husbandry, vegetative productions, developed in her great garden.

Let us suppose now that the future Man is an inch long, and just beginning to lose the quasi-larval form, and to stop up those terrible "scars" in the neck, the old fishy gill-clefts, the stigmas of ancestral lowliness. Along the line, where the skeleton is being developed, the cells are crowded, and soon these gatherings together of a special kind of protoplasmic centres become cartilage—hyaline cartilage, a substance that is somewhat elastic, and is like a solid sort of cheese. When this tissue has been formed by the deposit of a semi-solid substance between the cells, then we have a marvellous little model of the skeleton that is to be; as yet, however, there is no bone. Let us take that one little rod that is to be the thigh bone of the Man— as large as his flute—but now not more solid nor thicker than the filaments inside a flower. It is all hyaline cartilage at present, and the substance between the cells is in a very small quantity; yet the number of the centres of life—the protoplasmic masses—goes on increasing rapidly. This small, primary thigh-skeleton has this formative energy (*nisus formativus*) within it; but it must have an incessant, never-failing supply of food, and of fresh materials in the form of other species of protoplasmic cells, from without. The food, of course, is the plasmic or nourishing part of the blood, which is now developed, and for the containing and conveying of which proper vessels, and the heart itself, have been formed in connection with the other parts of the embryo. The fluid which is conveyed by, but escapes through, the vessels bathes every tissue, distilling itself like the dew of the morning upon and into every growing organ. The little masses of protoplasm that surround the young cartilaginous *femur*, or thigh-skeleton, are most important factors in the formation of this and the other organs of support. The cortical part forms a bark of fine fibres, like the "bast" or flax-layer in the stalk of the plant,—it is called the perichondrium—that which invests the cartilage. After a while, a most delicate ring of bone forms round the little rod of cartilage, between it and its bark of fibrous tissue; this is formed by the conversion of the innermost layer of that investment into bone-cells ("osteoblasts") by the deposit in them of phosphate of lime. And this takes place by a vegetative process, anterior to the time when proper animal functions are exercised. This little ring made of a paste, half earth and half protoplasm, grows larger, and becomes a sheath; then another, larger than it, is formed over it, and so on, sheath over sheath. Thus the bone grows *exogenously*, like the trunk of the oak, the bark of fibrous tissue yielding fresh and fresh materials; this fibrous layer now takes the name of "periosteum," or the covering of the bone. When the bone is scarcely one fourth the length it attains in the adult, the cartilage on which the bone was formed degenerates into a lower kind of tissue, and is partly absorbed; the cells that remain become red, and afterwards they undergo a further degradation into yellow marrow-cells. This degradation of the inner part of the rod is followed, after a time, by the absorption of the earlier rings of bony deposits, so that the cavity for the marrow grows larger and larger. All this is done without detriment to the strong pillar of bone; it is a living process, and although there is degeneration of tissue, it is not like the hollowing out of an old tree trunk. After a time the concentric growth of bony rings or tubes is supplemented by bony deposits in the top, the "trochanter," and the base of the bone; in the upper and lower parts, the cartilage itself becomes transformed into a new and more solid tissue, and the new bony substance, after a further lapse of time, becomes excavated into small galleries, so as to make the ends of the bone spongy. All the cartilage is not so transformed; at the end of the bone its remains are seen as a pad or buffer; so important in the pressure of the body during the motion of the joints. This account, meagre as it is, must serve us here as typical of what takes place when cartilage is transformed into bone; it will do duty equally well for the "femur" of all the other Mammals—with the exception of the Ornithorhynchus and Echidna.

The bones that form the roof and most of the side walls of the skull, and the outermost bones of the upper face, are formed by transformation of a mere web of fibrous tissue without any pre-existing cartilage. But the whole skeleton, with its various cartilages, bones, joints, ligamentous bands, and the like, is but the rougher and coarser part of this organic building.

The furniture of the upper chamber, its crypts, and cells, and galleries; the organs of special sense, with their labyrinthic passages, their fringed curtains, and carved porches; the tongue, doubly portcullised with the teeth and lips; and, indeed, the whole of this clay-compounded Man, is but a curiously modified repetition of all that we are familiar with in the Mammalia, generally, that seem so far below us.

I have, in the Ninth Lecture, commented upon the new connection of the three great branches of modern science. I may, in conclusion, remark further that those who are familiar with the English literature of the seventeenth century know that, during a few decades, the minds of thinking men became to a remarkable degree enfranchised and expanded in all that relates to the motions of the heavenly bodies, and, indeed, as to the idea of the Cosmos as a whole.

Another great and healthy growth of human thought took place at the latter end of the last, and the beginning of the present, century, when the fathers of modern Geology gave us their researches and deductions.

Of course, *time* was not excluded from the thoughts of the great discoverers of modern Astronomy; but Geologists, taking this one moving globe, the earth, for their field of labour, found a most invaluable approximative measure of secular periods in the study of the fossiliferous strata.

Modern Biology was germinant in the Palæontological department of Geology; it could not fail to come into active life when the days of its appointed time had come. These three great births of human thought—Astronomy, Geology, Biology—belong to us and to our children.

A

CATALOGUE

OF

STANDARD WORKS

PUBLISHED BY

CHARLES GRIFFIN & COMPANY.

	PAGE
I.—Religious Works, ...	1
II.—Scientific ,, . . .	6
III.—Educational ,, ...	18
IV.—Works in General Literature, . .	23

LONDON:

12 EXETER STREET, STRAND.

NOTICE.

New Issue of this Important Work—Enlarged, in part Re-written, and thoroughly Revised to date.

NINETEENTH EDITION, *Royal 8vo.* *Handsome Cloth,* 10s. 6d.

A DICTIONARY OF
DOMESTIC MEDICINE AND HOUSEHOLD SURGERY,

BY

SPENCER THOMSON, M.D., EDIN., L.R.C.S.,

REVISED, AND IN PART RE-WRITTEN, BY THE AUTHOR,

AND BY

JOHN CHARLES STEELE, M.D.,

OF GUY'S HOSPITAL.

With Appendix on the Management of the Sick-room, and many Hints for the Diet and Comfort of Invalids.

In its New Form, DR. SPENCER THOMSON'S "DICTIONARY OF DOMESTIC MEDICINE" fully sustains its reputation as the "Representative Book of the Medical Knowledge and Practice of the Day" applied to Domestic Requirements.

The most recent IMPROVEMENTS in the TREATMENT OF THE SICK—in APPLIANCES for the RELIEF OF PAIN—and in all matters connected with SANITATION, HYGIENE, and the MAINTENANCE of the GENERAL HEALTH—will be found in the New Issue in clear and full detail; the experience of the Editors in the Spheres of Private Practice and of Hospital Treatment respectively, combining to render the Dictionary perhaps the most thoroughly practical work of the kind in the English Language. Many new Engravings have been introduced—improved Diagrams of different parts of the Human Body, and Illustrations of the newest Medical, Surgical, and Sanitary Apparatus.

⁎ *All Directions given in such a form as to be readily and safely followed.*

FROM THE AUTHOR'S PREFATORY ADDRESS.

"Without entering upon that difficult ground which correct professional knowledge and educated judgment can alone permit to be safely trodden, there is a wide and extensive field for exertion, and for usefulness, open to the unprofessional, in the kindly offices of a *true* DOMESTIC MEDICINE, the timely help and solace of a simple HOUSEHOLD SURGERY, or, better still, in the watchful care more generally known as 'SANITARY PRECAUTION,' which tends rather to preserve health than to cure disease. 'The touch of a gentle hand' will not be less gentle because guided by knowledge, nor will the *safe* domestic remedies be less anxiously or carefully administered. Life may be saved, suffering may always be alleviated. Even to the resident in the midst of civilization, the 'KNOWLEDGE IS POWER,' to do good; to the settler and emigrant it is INVALUABLE."

"Dr. Thomson has fully succeeded in conveying to the public a vast amount of useful professional knowledge."—*Dublin Journal of Medical Science.*

"The amount of useful knowledge conveyed in this Work is surprising."—*Medical Times and Gazette.*

"WORTH ITS WEIGHT IN GOLD TO FAMILIES AND THE CLERGY."—*Oxford Herald.*

LONDON: CHARLES GRIFFIN & CO., EXETER STREET, STRAND.

CHARLES GRIFFIN & COMPANY'S
LIST OF PUBLICATIONS.

RELIGIOUS WORKS.

ALTAR OF THE HOUSEHOLD (The); a
Series of Prayers and Selections from the Scriptures, for Domestic Worship, for every Morning and Evening in the Year. By the Rev. Dr. HARRIS, assisted by eminent Contributors, with an Introduction by the Rev. W. LINDSAY ALEXANDER, D.D. *New Edition, entirely Revised.* Royal 4to, with Steel Frontispiece. Cloth, gilt edges, 22/.

Illustrated with a Series of First-class Engravings on Steel, 28/.

May also be had bound in the following styles: half-bound calf, marbled edges; and levant morocco, antique, gilt edges.

ANECDOTES (CYCLOPÆDIA OF RELIGIOUS AND
MORAL). With an Introductory Essay by the Rev. GEORGE CHEEVER, D.D. *Thirty-fourth Thousand.* Crown 8vo. Cloth, 3/6.

*** These Anecdotes relate to no trifling subjects; and they have been selected, not for amusement, but for instruction. By those engaged in the tuition of the young, they will be found highly useful.

BIBLE HISTORY (A Manual of). By the Rev.
J. WYCLIFFE GEDGE, Diocesan Inspector of Schools for Winchester. Small 8vo. Cloth, neat, 7d.

"This small but very comprehensive Manual is much more than a mere summary of Bible History."—*Church Sunday School Magazine.*

BUNYAN'S PILGRIM'S PROGRESS. With
Expository Lectures by the Rev. ROBERT MAGUIRE, Incumbent of St. Olave's, Southwark. With Steel Engravings. *Second Edition.* Imperial 8vo. Cloth, gilt, 10/6.

THE LARGE-TYPE BUNYAN.

BUNYAN'S PILGRIM'S PROGRESS. With
Life and Notes, Experimental and Practical, by WILLIAM MASON. Printed in large type, and Illustrated with full-page Woodcuts. *Twelfth Thousand.* Crown 8vo. Bevelled boards, gilt, and gilt edges, 3/6.

BUNYAN'S SELECT WORKS. With an
Original Sketch of the Author's Life and Times. Numerous Engravings. *New Edition.* Two vols., super-royal 8vo. Cloth, 36/.

CHRISTIAN YEAR (The): Thoughts in Verse

for the Sundays and Holy Days throughout the Year. With an original Memoir of the Rev. JOHN KEBLE, by W. TEMPLE, Portrait, and sixteen beautiful Engravings on Steel, after eminent Masters. In 4to. Handsome cloth, 12/6.

Unique walnut boards, . . . 21/.
Morocco antique, ... 25/.

ILLUSTRATIONS.

Morning	.. .. after H. HOWARD, R.A.	The Old Mansion .. after C. W. RADCLYFFE.
Sunset	,, G. BARRETT.	The Cathedral Choir ,, LEVAINT.
A Mountain Stream	,, C. BENTLEY.	Sunset (after CLAUDE),, G. BARRETT
A River Scene	,, C. W. RADCLYFFE.	Moonlight ,, HOFLAND.
A Mountain Lake	,, J. M. W. TURNER.	Pastoral Landscape ,, C. W. RADCLYFFE.
A Greek Temple	,, D. ROBERTS, R.A.	Halt in the Desert ,, D. ROBERTS, R.A.
A Village Church	,, C. W. RADCLYFFE.	Guardian Angels ,, H. HOWARD, R.A.
The Wayside Cross	,, TONY JOHANNOT.	The Church Gate ,, C. W. RADCLYFFE.

"An Edition *de luxe*, beautifully got up . . . admirably adapted for a gift-book."—*John Bull.*

CHRISTIAN YEAR (The): With Memoir of the

Author by W. TEMPLE, Portrait, and Eight Engravings on Steel, after eminent Masters. *New Edition.* Small 8vo, toned paper. Cloth gilt, 5/.

Morocco elegant, ... 10/6.
Malachite, ... 12/6.

*** The above are the only issues of the "Christian Year" with Memoir and Portrait of the Author. In ordering, Griffin's Editions should be specified.

COMMENTARIES ON THE HOLY SCRIPTURES.

HENRY (Matthew): The HOLY BIBLE.

With a Commentary and Explanatory Notes. *New Edition.* In 3 vols., super-royal 8vo. Strongly bound in cloth, 50/.

SCOTT (Rev. Thomas): A COMMEN-

TARY ON THE BIBLE; containing the Old and New Testaments according to the Authorised Version, with Practical Observations, copious Marginal References, Indices, &c. *New Edition.* In 3 vols., royal 4to. Cloth, 63/.

CRUDEN'S COMPLETE CONCORDANCE

TO THE OLD AND NEW TESTAMENTS AND THE BOOKS CALLED APOCRYPHAL. Edited and Corrected by WILLIAM YOUNGMAN. With fine Portrait of CRUDEN. *New Edition.* Imperial 8vo. Cloth, handsome gilt top, 7/6.

DICK (Thos., LL.D.): CELESTIAL

SCENERY; or, The Wonders of the Planetary System Displayed. This Work is intended for general readers, presenting to their view, in an attractive manner, sublime objects of contemplation. Illustrated. *New Edition.* Crown 8vo, toned paper. Handsomely bound, gilt edges, 5/.

DICK (Dr.): CHRISTIAN PHILOSOPHER

(The); or, The Connection of Science and Philosophy with Religion. Revised and enlarged. Illustrated with 150 Engravings on Wood. *Twenty-eighth Edition.* Crown 8vo, toned paper. Handsomely bound, with gilt edges, 5/.

STANDARD BIBLICAL WORKS

BY

THE REV. JOHN EADIE, D.D., LL.D.,

Late a Member of the New Testament Revision Company.

This SERIES has been prepared to afford sound and necessary aid to the Reader of Holy Scripture. The VOLUMES comprised in it form in themselves a COMPLETE LIBRARY OF REFERENCE. The number of Copies already issued greatly exceeds A QUARTER OF A MILLION.

I. EADIE (Rev. Prof.): BIBLICAL CYCLO-
PÆDIA (A); or, Dictionary of Eastern Antiquities, Geography, and Natural History, illustrative of the Old and New Testaments. With Maps, many Engravings, and Lithographed Facsimile of the Moabite Stone. Large post 8vo, 700 pages. *Twenty-third Edition.* Handsome cloth, 7/6.

    Half-bound, calf, ... 10/6.
    Morocco antique, gilt edges, . . 16/.

"By far the best Bible Dictionary for general use."—*Clerical Journal.*

II. EADIE (Rev. Prof.): CRUDEN'S CON-
CORDANCE TO THE HOLY SCRIPTURES. With Portrait on Steel, and Introduction by the Rev. Dr. KING. Post 8vo. *Forty-seventh Edition.* Handsome cloth, ... 3/6.
    Half-bound, calf, . . . 6/6.
    Full calf, gilt edges, . . . 8/6.
    Full morocco, gilt edges, . . 10/6.

* Dr. EADIE'S has long and deservedly borne the reputation of being the COMPLETEST and BEST CONCORDANCE extant.

III. EADIE (Rev. Prof.): CLASSIFIED BIBLE
(The). An Analytical Concordance. Illustrated by Maps. Large Post 8vo. *Sixth Edition.* Handsome cloth, . . . 8/6.
    Full morocco, antique, . . . 17/.

"We have only to add our unqualified commendation of a work of real excellence to every Biblical student."—*Christian Times.*

IV. EADIE (Rev. Prof.): ECCLESIASTICAL
CYCLOPÆDIA (The). A Dictionary of Christian Antiquities, and of the History of the Christian Church. By the Rev. Professor EADIE, assisted by numerous Contributors. Large Post 8vo. *Sixth Edition.* Handsome cloth, ... 8/6.
    Full morocco, antique, . . . 17/.

"The ECCLESIASTICAL CYCLOPÆDIA will prove acceptable both to the clergy and laity of Great Britain. A great body of useful information will be found in it."—*Athenæum.*

V. EADIE (Rev. Prof.): A DICTIONARY OF
THE HOLY BIBLE; for the use of Young People. With Map and Illustrations. Small 8vo. *Thirty-sixth Thousand.* Cloth, elegant, 2/6.
    Full morocco, gilt edges, . . . 7/6.

FOSTER (Charles): THE STORY OF THE

BIBLE, from Genesis to Revelation—including the Historical Connection between the Old and New Testaments. Told in Simple Language. Large Post 8vo, with Maps and over 250 Engravings (many of them Full-page, after the Drawings of Professor CARL SCHÖNHERR and others), illustrative of the Bible Narrative, and of Eastern Manners and Customs. *Now Ready. Third Thousand.*

Home and School Edition, cloth elegant, . . 6/.
Prize and Presentation Edition, beautifully gilt. . 7/6.

OPINIONS OF THE PRESS.

"A book which, once taken up, is not easily laid down. When the volume is opened, we are fairly caught. Not to speak of the well-executed wood engravings, which will each tell its story, we find a simple version of the main portions of the Bible, all that may most profitably be included in a work intended at once to instruct and charm the young—a version couched in the simplest, purest, most idiomatic English, and executed throughout with good taste, and in the most reverential spirit. *The work needs only to be known to make its way into families,* and it will (at any rate, it *ought* to) become a favourite Manual in Sunday Schools."—*Scotsman.*

"A HOUSEHOLD TREASURE."—*Western Morning News.*

"This attractive and handsome volume . . . written in a simple and transparent style. . . . Mr. Foster's explanations and comments are MODELS OF TEACHING."—*Freeman.*

"This large and handsome volume, abounding in Illustrations, is just what is wanted. . . The STORY is very beautifully and reverently told."—*Glasgow News.*

"There could be few better Presentation Books than this handsome volume."—*Daily Review.*

"This elegant volume will prove a valuable adjunct in the Home Circle and Sunday Class."—*Western Daily Mercury.*

"WILL ACCOMPLISH A GOOD WORK."—*Sunday School Chronicle.*

"In this beautiful volume no more of comment is indulged in than is necessary to the elucidation of the text. Everything approaching Sectarian narrowness is carefully eschewed."—*Methodist Magazine.*

"This simple and impressive Narrative . . . succeeds thoroughly in rivetting the attention of children; . . . admirably adapted for reading in the Home Circle."—*Daily Chronicle.*

"The HISTORICAL SKETCH connecting the Old and New Testaments is a very good idea; it is a common fault to look on these as distinct histories, instead of as parts of *one grand whole.*"—*Christian.*

"Sunday School Teachers and Heads of Families will best know how to value this handsome volume."—*Northern Whig.*

KITTO (John, D.D., F.S.A.): THE HOLY

LAND : The Mountains, Valleys, and Rivers of the Holy Land; being the Physical Geography of Palestine. With eight full-page Illustrations. *Eleventh Thousand. New Edition.* Fcap 8vo. Cloth, 2/6.

⁎ Contains within a small compass a body of most interesting and valuable information

—— THE PICTORIAL SUNDAY BOOK:

Containing nearly two thousand Illustrations on Steel and Wood, and a Series of Maps. *Seventy-third Thousand.* Folio. Cloth, gilt, 30/.

PALEY (Archdeacon): NATURAL THEOLOGY.

The Evidences of the Existence and the Attributes of the Deity. With Illustrative Notes and Dissertations, by HENRY, Lord BROUGHAM, and Sir CHARLES BELL. Many Engravings. One vol., 16mo. Cloth, 4/.

—— With Lord BROUGHAM'S NOTES AND DIALOGUES ON INSTINCT. Many Illustrations. Three vols., 16mo. Cloth, 7/6.

"When Lord Brougham's eloquence in the Senate shall have passed away, and his services as a statesman shall exist only in the free institutions which they have helped to secure, his discourse on Natural Theology will continue to inculcate imperishable truths, and fit the mind for the higher revelations which these truths are destined to foreshadow and confirm."—*Edinburgh Review.*

RAGG (Rev. Thomas): CREATION'S TESTIMONY TO ITS GOD:

the Accordance of Science, Philosophy, and Revelation. *Thirteenth Edition.* Large crown 8vo. Handsome cloth, bevelled boards, 5/.

"We are not a little pleased again to meet with the author of this volume in the new edition of his far-famed work. Mr. Ragg is one of the few original writers of our time to whom justice is being done." *British Standard.*

⁂ This work has been pronounced "The Book of the Age," "The best popular Text-Book of the Sciences," and "The only complete Manual of Religious Evidence, Natural and Revealed."

RELIGIONS OF THE WORLD (The): Being

Confessions of Faith contributed by Eminent Members of every Denomination of Christians, &c. of Mahometanism, Parseeism, Brahminism, Mormonism, &c., &c., with a Harmony of the Christian Confessions of Faith by a Member of the Evangelical Alliance. Crown 8vo. Cloth bevelled, 3/6.

⁂ In this volume, each denomination, through some leading member, has expressed its own opinions. There is no book in the language on the same plan. All other works on the subject, being written by one individual, are necessarily one-sided, incomplete, and unauthentic.

SOUTHGATE (Henry): SUGGESTIVE THOUGHTS ON RELIGIOUS SUBJECTS. (See page 31.)

SOUTHGATE (Mrs. Henry): THE CHRISTIAN LIFE:

Thoughts in Prose and Verse from Five hundred of the Best Writers of all Ages. Selected and Arranged for Every Day in the Year. Small 8vo. With Red Lines and unique Initial Letters on each page. Cloth Elegant, 5/. Morocco, 10/6.

"A volume as handsome as it is intrinsically valuable."—*Scotsman.*
"The readings are excellent."—*Record.*
"A library in itself."—*Northern Whig.*

WORDS AND WORKS OF OUR BLESSED LORD:

and their Lessons for Daily Life. Two Vols. in One. Foolscap, 8vo. Cloth, gilt edges, 6/.

SCIENTIFIC WORKS.

MEDICAL WORKS

By WILLIAM AITKEN, M.D., Edin., F.R.S.,

PROFESSOR OF PATHOLOGY IN THE ARMY MEDICAL SCHOOL; EXAMINER IN MEDICINE FOR THE MILITARY MEDICAL SERVICES OF THE QUEEN; FELLOW OF THE SANITARY INSTITUTE OF GREAT BRITAIN; CORRESPONDING MEMBER OF THE ROYAL IMPERIAL SOCIETY OF PHYSICIANS OF VIENNA; AND OF THE SOCIETY OF MEDICINE AND NATURAL HISTORY OF DRESDEN.

Now Ready. Seventh Edition.

The SCIENCE and PRACTICE of MEDICINE.

In Two Volumes, Royal 8vo., cloth. Illustrated by numerous Engravings on Wood, and a Map of the Geographical Distribution of Diseases. To a great extent Rewritten; Enlarged, Remodelled, and Carefully Revised throughout, 42/.

In reference to the Seventh Edition of this important Work, the Publishers would only remark, that no labour or expense has been spared to sustain its well-known reputation as "The Representative Book of the Medical Science and Practice of the Day." Among the More Important Features of the New Edition, the subject of DISEASES OF THE BRAIN AND NERVOUS SYSTEM *may be specially mentioned.*

Opinions of the Press.

"The work is an admirable one, and adapted to the requirements of the Student, Professor, and Practitioner of Medicine. . . . The reader will find a large amount of information not to be met with in other books, epitomised for him in this. We know of no work that contains so much, or such full and varied information on all subjects connected with the Science and Practice of Medicine."—*Lancet.*

"Excellent from the beginning, and improved in each successive issue, Dr. Aitken's GREAT and STANDARD WORK has now, with vast and judicious labour, been brought abreast of every recent advance in scientific medicine and the healing art, and affords to the Student and Practitioner a store of knowledge and guidance of altogether inestimable value. . . . The first 530 Pages of the Second Volume would, if printed separately, form perhaps the best text-book in our language for the student of Neurology and Insanity. A masterly and philosophical review, characterised by the precision of the specialist, and the breadth of the catholic physician, is presented in these pages of the varied phenomena connected with morbid conditions of the nervous system in their relation to anatomical structure, chemical composition, physiological uses, and pathological changes. . . . A classical work which does honour to British Medicine, and is a compendium of sound knowledge."—*Extract from Review in "Brain," by J. Crichton-Browne, M.D., F.R.S., Lord Chancellor's Visitor in Lunacy.*

"The SEVENTH EDITION of this important Text-Book fully maintains its reputation. . . . Dr. Aitken is indefatigable in his efforts. . . . The section on DISEASES of the BRAIN and NERVOUS SYSTEM is completely remodelled, so as to include all the most recent researches, which in this department have been not less important than they are numerous."—*British Medical Journal.*

"The STANDARD TEXT-BOOK in the English Language. . . . There is, perhaps, no work more indispensable for the Practitioner and Student."—*Edin. Medical Journal.*

"In its system, in its scope, and in its method of dealing with the subjects treated of, this work differs from all other Text-Books on the Science of Medicine in the English language."—*Medical Times and Gazette.*

"The extraordinary merit of Dr. Aitken's work. . . . The author has unquestionably performed a service to the profession of the most valuable kind."—*Practitioner.*

"Altogether this voluminous treatise is a credit to its Author, its Publisher, and to English Physic. . . . Affords an admirable and honest digest of the opinions and practice of the day. . . . Commends itself to us for sterling value, width of retrospect, and fairness of representation."—*Medico-Chirurgical Review.*

"Diseases are here described which have hitherto found no place in any English systematic work."—*Westminster Review.*

"We can say with perfect confidence, that no medical man in India should be without Dr. Aitken's 'Science and Practice of Medicine.' The article on Cholera is by far the most complete, judicious, and learned summary of our knowledge respecting this disease which has yet appeared."—*Indian Medical Gazette.*

PROF. AITKEN'S WORKS—*(Continued)*.

———— OUTLINES OF THE SCIENCE AND
PRACTICE OF MEDICINE. A Text-Book for Students. *Second Edition.* Crown 8vo, 12/6.

"Students preparing for examinations will hail it as a perfect godsend for its conciseness."—*Athenæum*.

"Well-digested, clear, and well-written, the work of a man conversant with every detail of his subject, and a thorough master of the art of teaching."—*British Medical Journal*.

"In respect of both the matter contained, and the manner in which it is conveyed, our examination has convinced us that nothing could be better. . . . We know of no summary of the use of Electricity as a means of diagnosis, equal to that contained in the Section on Diseases of the Nervous System." *Medico-Chirurgical Review*.

———— THE GROWTH OF THE RECRUIT,
and the Young Soldier, with a view to the Selection of "Growing Lads" and their Training, 2/6.

ANSTED (Prof., M.A., F.R.S.) : NATURAL
HISTORY OF THE INANIMATE CREATION, recorded in the Structure of the Earth, the Plants of the Field, and the Atmospheric Phenomena. With numerous Illustrations. Large post 8vo. Cloth, 8/6.

BAIRD (W., M.D., F.L.S., late of the Brit. Mus.) :
THE STUDENT'S NATURAL HISTORY; a Dictionary of the Natural Sciences : Botany, Conchology, Entomology, Geology, Mineralogy, Palæontology, and Zoology. With a Zoological Chart, and over 250 Illustrations. Demy 8vo. Cloth gilt, 10/6.

"The work is a very useful one, and will contribute, by its cheapness and comprehensiveness, to foster the extending taste for Natural Science."—*Westminster Review*.

BROWNE (Walter R., M.A., M. Inst. C.E.,
F.G.S., late Fellow of Trinity College, Cambridge) :

THE STUDENT'S MECHANICS : An Introduction to the Study of Force and motion. With Diagrams. Crown 8vo. Cloth, 4/6.

Contents.

I. FIRST PRINCIPLES.
II. STATICS.
III. KINEMATICS.
IV. DYNAMICS.
V. AXIOMS, DEFINITIONS, AND LAWS.
VI. EXAMPLES WORKED & UNWORKED.

"Clear in style and practical in method, "THE STUDENT'S MECHANICS," is cordially to be recommended from all points of view. . . . Will be of great value to Students desirous to gain full knowledge."—*Athenæum*.

"The merits of the work are especially conspicuous in its clearness and brevity . . . deserves the attention of all who have to teach or learn the elements of Mechanics . . . An excellent conception."—*Westminster Review*.

———— FOUNDATIONS OF MECHANICS.
Papers reprinted from the *Engineer*. In crown 8vo, 1/.

———— FUEL AND WATER: A Manual for
Users of Steam and Water. By Prof. SCHWACKHÖFER and W. R. BROWNE, M.A. (See p. 16.)

WORKS by A. WYNTER BLYTH, M.R.C.S., F.C.S.,
Public Analyst for the County of Devon, and Medical Officer of Health for
St. Marylebone.

HYGIÈNE AND PUBLIC HEALTH (a Dictionary of): embracing the following subjects:—

I.—SANITARY CHEMISTRY: the Composition and Dietetic Value of Foods, with the Detection of Adulterations.
II.—SANITARY ENGINEERING: Sewage, Drainage, Storage of Water, Ventilation, Warming, &c.
III.—SANITARY LEGISLATION: the whole of the PUBLIC HEALTH ACT, together with portions of other Sanitary Statutes (without alteration or abridgment, save in a few unimportant instances), in a form admitting of easy and rapid Reference.
IV.—EPIDEMIC AND EPIZOOTIC DISEASES: their History and Propagation, with the Measures for Disinfection.
V —HYGIÈNE—MILITARY, NAVAL, PRIVATE, PUBLIC, SCHOOL.

Royal 8vo, 672 pp., cloth, with Map and 140 Illustrations, 28/.

Opinions of the Press.

"Excellently done . . . the articles are brief, but comprehensive. We have tested the book, and can therefore recommend Mr. Blyth's Dictionary with confidence."—*Westminster Review.*

"A very important Treatise . . . an examination of its contents satisfies us that it is a work which should be highly appreciated."—*Medico-Chirurgical Review.*

"A work that must have entailed a vast amount of labour and research. . . . Will become a STANDARD WORK IN HYGIENE AND PUBLIC HEALTH."—*Medical Times and Gazette.*

"Contains a great mass of information of easy reference."—*Sanitary Record.*

RE-ISSUE OF BLYTH'S "PRACTICAL CHEMISTRY."

In 2 Vols. Crown 8vo.

Vol. I.—FOODS: THEIR COMPOSITION AND ANALYSIS. Price 16/.

General Contents.

History of Adulteration - Legislation, Past and Present--Apparatus useful to the Food Analyst—"Ash"—Sugar—Confectionery Honey—Treacle—Jams and Preserved Fruits Starches—Wheaten-Flour—Bread—Oats— Barley—Rye—Rice—Maize—Millet—Potato–Peas—Chinese Peas—Lentils—Beans—MILK—Cream Butter—Cheese—Tea—Coffee—Cocoa and Chocolate—Alcohol—Brandy Rum—Whisky—Gin—Arrack—Liqueurs—Beer—Wine—Vinegar—Lemon and Lime Juice—Mustard—Pepper—Sweet and Bitter Almond—Annatto—Olive Oil—Water. *Appendix:* Text of English and American Adulteration Acts.

In Crown 8vo, cloth, with Elaborate Tables, Folding Litho-Plate, and Photographic Frontispiece.

Vol. II.—POISONS: THEIR EFFECTS AND DETECTION. Price 16/.

General Contents.

Historical Introduction—Statistics--General Methods of Procedure—Life Tests—Special Apparatus-- Classification: I.—ORGANIC POISONS: (*a.*) Sulphuric, Hydrochloric, and Nitric Acids, Potash, Soda, Ammonia, &c.; (*b.*) Petroleum, Benzene, Camphor, Alcohols, Chloroform, Carbolic Acid, Prussic Acid, Phosphorus, &c.; (*c.*) Hemlock, Nicotine, Opium, Strychnine, Aconite, Atropine, Digitalis, &c.; (*d.*) Poisons derived from Animal Substances; (*e.*) The Oxalic Acid Group. II.—INORGANIC POISONS: Arsenic, Antimony, Lead, Copper, Bismuth, Silver, Mercury, Zinc, Nickel Iron, Chromium, Alkaline Earths, &c. *Appendix:* A. Examination of Blood and Blood-Spots. B. *Hints for Emergencies:* Treatment—Antidotes.

"Will be used by every Analyst."—*Lancet.*

"Full of great interest . . . The method of treatment excellent . . . Gives just that amount of information which those able to appreciate it will desire."—*Westminster Review.*

"STANDS UNRIVALLED for completeness of information . . . A really 'practical' work for the guidance of practical men."—*Sanitary Record.*

"The whole work is full of useful practical information."—*Chemical News.*

THE CIRCLE OF THE SCIENCES: A
SERIES OF POPULAR TREATISES ON THE NATURAL AND PHYSICAL SCIENCES, AND THEIR APPLICATIONS, by Professors OWEN, ANSTED, YOUNG, and TENNANT; Drs. LATHAM, EDWARD SMITH, SCOFFERN, BUSHNAN, and BRONNER; Messrs. MITCHELL, TWISDEN, DALLAS, GORE, IMRAY, MARTIN, SPARLING, and others. Complete in nine volumes, illustrated with many thousand Engravings on Wood. Crown 8vo. Cloth lettered. Each vol., 5/.

VOL. 1.—ORGANIC NATURE.—Part I. Animal and Vegetable Physiology; the Skeleton and the Teeth; Varieties of the Human Race.
VOL. 2.—ORGANIC NATURE.—Part II. Structural and Systematic Botany—Invertebrated Animals.
VOL. 3.—ORGANIC NATURE.—Part III. Vertebrated Animals.
VOL. 4.—INORGANIC NATURE. Geology and Physical Geography; Crystallography; Mineralogy; Meteorology, and Atmospheric Phenomena.
VOL. 5.—NAVIGATION: PRACTICAL AND NAUTICAL ASTRONOMY.
VOL. 6.—ELEMENTARY CHEMISTRY.
VOL. 7.—PRACTICAL CHEMISTRY.—Electro-Metallurgy; Photography; Chemistry of Food; and Artificial Light.
VOL. 8.—MATHEMATICAL SCIENCE.—Arithmetic; Algebra; Plane Geometry; Logarithms; Plane and Spherical Trigonometry; Mensuration and Practical Geometry, with use of Instruments.
VOL. 9.—MECHANICAL PHILOSOPHY.—Statics; Dynamics; Hydrostatics; Pneumatics; Practical Mechanics; and the Steam Engine.

IN SEPARATE TREATISES. Cloth.

1. ANSTED's Geology and Physical Geography, . . . 2/6.
2. BREEM's Practical Astronomy, ... 2/6.
3. BRONNER and SCOFFERN's Chemistry of Food and Diet. . 1/6.
4. BUSHNAN's Physiology of Animal and Vegetable Life, . 1/6.
5. GORE's Theory and Practice of Electro-Deposition, . . 1/6.
6. IMRAY's Practical Mechanics, ... 1/6.
7. JARDINE's Practical Geometry, ... 1/.
8. LATHAM's Varieties of the Human Species, . . . 1/6.
9. MITCHELL and TENNANT's Crystallography and Mineralogy, 3/.
10. MITCHELL's Properties of Matter and Elementary Statics, 1/6.
11. OWEN's Principal Forms of the Skeleton and the Teeth, . 1/6.
12. SCOFFERN's Chemistry of Heat, Light, and Electricity, . 3/.
13. SCOFFERN's Chemistry of the Inorganic Bodies, . . 3/.
14. SCOFFERN's Chemistry of Artificial Light, . . . 1/6.
15. SCOFFERN and LOWE's Practical Meteorology, . . 1/6.
16. SMITH's Introduction to Botany: Structural and Systematic, 2/.
17. TWISDEN's Plane and Spherical Trigonometry, . . 1/6.
18. TWISDEN on Logarithms, ... 1/.
19. YOUNG's Elements of Algebra, ... 1/.
20. YOUNG's Solutions of Questions in Algebra, . . . 1/.
21. YOUNG's Navigation and Nautical Astronomy, . . . 2/6.
22. YOUNG's Plane Geometry, ... 1/6.
23. YOUNG's Simple Arithmetic, ... 1/.
24. YOUNG's Elementary Dynamics, ... 1/6.

DALLAS (W. S., F.L.S.):

A POPULAR HISTORY OF THE ANIMAL CREATION: The Habits, Structure, and Classification of Animals. With coloured Frontispiece and many hundred Illustrations. *New Edition.* Crown 8vo. Cloth, 8/6.

DOUGLAS (John Christie, Mem. Soc. Tel. Eng., East India Govt. Telegraph Department, &c.):

A MANUAL OF TELEGRAPH CONSTRUCTION. For the use of Telegraph Engineers and others. With numerous Diagrams. Crown 8vo. Cloth, bevelled, 15/.

** *Second Edition. Published with the approval of the Director-General of Telegraphs in India.*

GENERAL CONTENTS.

PART I.—GENERAL PRINCIPLES OF STRENGTH AND STABILITY: with the Strength of Materials.

PART II.—PROPERTIES AND APPLICATIONS OF MATERIALS, WITH SPECIFICATIONS.

PART III.—TELEGRAPH CONSTRUCTION, MAINTENANCE, AND ORGANISATION, treating of the Application of the Information conveyed in Parts I. and II. to the case of Combined Structures; including the Construction of Overground, Subterranean, and Subaqueous Lines; Office Fittings; Estimating; Organisation, &c.

"Mr. Douglas deserves the thanks of Telegraphic Engineers for the excellent 'Manual' now before us . . . he has ably supplied an existing want . . . the subject is treated with great clearness and judgment . . . good practical information, given in a clear, terse style."—*Engineering.*

"The amount of information given is such as to render this volume a most useful guide to any one who may be engaged in any branch of Electric Telegraph Engineering."—*Athenæum.*

"Calculated to be of great service to Telegraphic Engineers." —*Iron.*

DUPRÉ (Auguste, Ph. D., F.R.S., Prof. of Chemistry at the Westminster Hospital) and HAKE (H. W., Ph. D., F.C.S., of Queenwood College):

A MANUAL OF CHEMISTRY, Organic and Inorganic, for the use of Students. *(In preparation).*

GRIFFIN (John Joseph, F.C.S.):

CHEMICAL RECREATIONS: A Popular Manual of Experimental Chemistry. With 540 Engravings of Apparatus. *Tenth Edition.* Crown 4to. Cloth.

Part I.—Elementary Chemistry, 2/.

Part II.—The Chemistry of the Non-Metallic Elements, including a Comprehensive Course of Class Experiments, 10/6.

Or, complete in one volume, cloth, gilt top, . . 12/6.

GURDEN (Richard Lloyd, Authorised Surveyor
for the Governments of New South Wales and Victoria):

TRAVERSE TABLES: computed to Four Places Decimals for every Minute of Angle up to 100 of Distance. For the use of Surveyors and Engineers. In folio, strongly half-bound, 30/.

*** *Published with Concurrence of the Surveyors-General for New South Wales and Victoria.*

"Mr. GURDEN is to be thanked for the extraordinary labour which he has bestowed on facilitating the work of the Surveyor. . . . An almost unexampled instance of professional and literary industry."—*Athenæum*.

"Those who have experience in exact SURVEY-WORK will best know how to appreciate the enormous amount of labour represented by this valuable book. The computations enable the user to ascertain the sines and cosines for a distance of twelve miles to within half an inch, and this BY REFERENCE TO BUT ONE TABLE, in place of the usual Fifteen minute computations required. This alone is evidence of the assistance which the Tables ensure to every user, and as every Surveyor in active practice has felt the want of such assistance, few knowing of their publication will remain without them."—*Engineer*.

"We cannot sufficiently admire the heroic patience of the author, who, in order to prevent error, calculated each result by two different modes, and, before the work was finally placed in the Printers' hands, repeated the operation for a third time, on revising the proofs."—*Engineering*.

"Up to the present time, no Tables for the use of Surveyors have been prepared, which, in minuteness of detail, can be compared with those compiled by Mr Gurden. . . . With the aid of this book, *the toil of calculation is reduced to a minimum;* and not only is time saved, but the risk of error is avoided. Mr. Gurden's book has but to be known, and no Engineer's or Architect's office will be without a copy."—*Architect*.

"A valuable acquisition to those employed in extensive surveys."—*Building News*.

"These Tables are Characterised by ABSOLUTE SIMPLICITY; the saving of time effected by their use is most material. . . . Every one connected with Engineering or Survey should be made aware of the existence of this elaborate and useful set of Tables."—*Builder*.

JAMIESON (Andrew, C.E., F.R.S.E.):

STEAM AND THE STEAM ENGINE (A Manual of) for the use of Students preparing for Government and other Competitive Examinations. With Numerous Diagrams. Crown 8vo. (*In preparation*.)

JAMIESON (Andrew, C.E.), and MUNRO (John, C.E.):

A POCKET-BOOK OF ELECTRICAL RULES AND TABLES. —(*See* Munro, John.)

LEARED (Arthur, M.D., F.R.C.P., late Senior
Physician to the Great Northern Hospital:

IMPERFECT DIGESTION: Its Causes and Treatment. Post 8vo. *Seventh Edition.* Cloth, 4/6.

"It now constitutes about the best work on the subject."—*Lancet*.

"Dr. Leared has treated a most important subject in a practical spirit and popular manner."—*Medical Times and Gazette*.

LINN (S.H., M.D., D.D.S., Dentist to the Imperial
Medico-Chirurgical Academy of St. Petersburg):

THE TEETH: How to preserve them and prevent their Decay. A Popular Treatise on the Diseases and the Care of the Teeth. With Plates and Diagrams. Crown 8vo. Cloth, 2/6.

"Everyone who values his teeth—(and who does not?)—should study this practical little book."

"Many important truths on the preservation of the teeth and the irregularity of children's teeth are here set forth; and on the subject of artificial teeth there is sound advice, which most of us may sooner or later be glad of."—*Medical Times and Gazette.*

"Contains much useful information and excellent advice." *Leeds Mercury.*

"Deserves to be widely read." *Northern Whig.*

"We heartily recommend the treatise." —*John Bull.*

LONGMORE (Surgeon-General, C.B., Q.H.S.,
F.R.C.S., &c., Professor of Military Surgery in the Army Medical School):

THE SANITARY CONTRASTS OF THE CRIMEAN WAR. Demy 8vo. Cloth limp, 1/6.

"A most valuable contribution to Military Medicine."—*British Medical Journal.*

"A most concise and interesting Review."—*Lancet.*

M'NAB (W. Ramsay, M.D., F.L.S., Professor of
Botany at the Royal College of Science, Dublin):

A MANUAL OF BOTANY, Structural and Systematic, for the use of Students. (*In preparation.*)

MOFFITT (Staff-Assistant-Surgeon A., late of the
Royal Victoria Hospital, Netley):

A MANUAL OF INSTRUCTION FOR ATTENDANTS ON THE SICK AND WOUNDED IN WAR. With numerous Illustrations. Post 8vo. Cloth, 5/.

** *Published under the sanction of the National Society for Aid to the Sick and Wounded in War.*

"A work by a practical and experienced author. After an explicit chapter on the Anatomy of the Human Body, directions are given concerning bandaging, dressing of sores, wounds, &c., assistance to wounded on field of action, stretchers, mule litters, ambulance, transport, &c. All Dr. Moffitt's instructions are assisted by well-executed illustrations."—*Public Opinion.*

MUNRO (John, C.E.) and JAMIESON
(Andrew, C.E., F.R.S.E.):

A POCKET-BOOK OF ELECTRICAL RULES AND TABLES, for the use of Electricians and Engineers. Royal 32mo. (*At press.*)

NAPIER (James, F.R.S.E., F.C.S.):

A MANUAL OF THE ART OF DYEING AND DYEING RECEIPTS. Illustrated by Diagrams and Numerous Specimens of Dyed Cotton, Silk, and Woollen Fabrics. Demy 8vo. *Third Edition, thoroughly revised and greatly enlarged.* Cloth, 21/.

General Contents.

I. HEAT AND LIGHT.
II. A CONCISE SYSTEM OF CHEMISTRY, with special reference to Dyeing.
III. MORDANTS AND ALTERANTS.

IV. VEGETABLE MATTERS in use in the Dye-house.
V. ANIMAL DYES.
VI. COAL-TAR COLOURS.

APPENDIX—RECEIPTS FOR MANIPULATION.

"The numerous Dyeing Receipts and the Chemical Information furnished will be exceedingly valuable to the Practical Dyer. . . . A Manual of necessary reference to all those who wish to master their trade, and keep pace with the scientific discoveries of the time."—*Journal of Applied Science.*

"In this work Mr. Napier has done good service . . . being a Practical Dyer himself, he knows the wants of his *Confrères* . . . the Article on Water is a very valuable one to the Practical Dyer, enabling him readily to detect impurities, and correct their action. The Article on Indigo is very exhaustive . . . the Dyeing Receipts are very numerous, and well illustrated." - *Textile Manufacturer.*

—— A MANUAL OF ELECTRO-METAL-LURGY.

With numerous Illustrations. Crown 8vo. Cloth. *Fifth Edition, revised and enlarged,* 7/6.

General Contents.

I. HISTORY of the ART.
II. DESCRIPTION of GALVANIC BATTERIES and their RESPECTIVE PECULIARITIES.
III. ELECTROTYPE PROCESSES.
IV. BRONZING.
V. MISCELLANEOUS APPLICATIONS of the PROCESS of COATING with COPPER.

VI. DEPOSITION of METALS upon one another.
VII. ELECTRO-PLATING.
VIII. ELECTRO-GILDING.
IX. RESULTS of EXPERIMENTS on the DEPOSITIONS of other METALS as COATINGS.
X. THEORETICAL OBSERVATIONS.

"A work that has become an established authority on Electro-Metallurgy, an art which has been of immense use to the Manufacturer in *economising the quantity of the precious metals absorbed, and in extending the sale of Art Manufactures.* . . . We can heartily commend the work as a valuable handbook on the subject on which it treats."—*Journal of Applied Science.*

"The fact of Mr. Napier's Treatise having reached a FIFTH EDITION is good evidence of an appreciation of the Author's mode of treating his subject. . . . A very useful and practical little Manual."—*Iron.*

"The Fifth Edition has all the advantages of a new work, and of a proved and tried friend. Mr. Napier is well known for the carefulness and accuracy with which he writes. . . . There is a thoroughness in the handling of the subject which is far from general in these days. . . . The work is one of those which, besides supplying first-class information, are calculated to inspire invention."—*Jeweller and Watchmaker.*

PHILLIPS (John, M.A., F.R.S., late Professor of Geology in the University of Oxford):

A MANUAL OF GEOLOGY: Theoretical and Practical. Edited by ROBERT ETHERIDGE, F.R.S., Palæontologist to the Geological Survey of Great Britain, Past-President of the Geological Society; and HARRY GOVIER SEELEY, F.R.S., Professor of Geography in King's College, London. With numerous Tables, Sections, and Figures of Characteristic Fossils. *In Preparation.* Demy 8vo, *Third Edition:* Thoroughly Revised and Augmented.

PHILLIPS (J. Arthur, M. Inst. C.E., F.C.S., F.G.S.,
Ancien Élève de l'École des Mines, Paris):

ELEMENTS OF METALLURGY: a Practical Treatise on the Art of Extracting Metals from their Ores. With over 200 Illustrations, many of which have been reduced from Working Drawings. Royal 8vo., 764 pages, cloth, 34/.

General Contents.

I.—A TREATISE on FUELS and REFRACTORY MATERIALS.

II.—A Description of the principal METALLIFEROUS MINERALS, with their DISTRIBUTION.

III.—STATISTICS of the amount of each METAL annually produced throughout the World, obtained from official sources, or, where this has not been practicable, from authentic private information.

IV.—The METHODS of ASSAYING the different ORES, together with the PROCESSES of METALLURGICAL TREATMENT, comprising:

Refractory Materials.	Antimony.	Iron.
Fire-Clays.	Arsenic.	Cobalt.
Fuels, &c.	Zinc.	Nickel.
Alluminium.	Mercury.	Silver.
Copper.	Bismuth.	Gold.
Tin.	Lead.	Platinum.

"'Elements of Metallurgy' possesses intrinsic merits of the highest degree. Such a work is precisely wanted by the great majority of students and practical workers, and its very compactness is in itself a first-rate recommendation. . . . In our opinion, the BEST WORK EVER WRITTEN ON THE SUBJECT with a view to its practical treatment."—*Westminster Review.*

"In this most useful and handsome volume, Mr. Phillips has condensed a large amount of valuable practical knowledge. . . . We have not only the results of scientific inquiry most cautiously set forth, but the experiences of a thoroughly practical man very clearly given."—*Athenæum.*

"For twenty years the learned author, who might well have retired with honour on account of his acknowledged success and high character as an authority in Metallurgy, has been making notes, both as a Mining Engineer and a practical Metallurgist, and devoting the most valuable portion of his time to the accumulation of materials for this his Masterpiece."—*Colliery Guardian.*

"The VALUE OF THIS WORK IS ALMOST INESTIMABLE. There can be no question that the amount of time and labour bestowed upon it is enormous. . . . There is certainly no Metallurgical Treatise in the language calculated to prove of such general utility to the Student really seeking sound practical information upon the subject, and none which gives greater evidence of the extensive metallurgical knowledge of its Author."—*Mining Journal.*

PORTER (Surgeon-Major J. H., Late Assistant
Professor of Military Surgery in the Army Medical School, and Hon. Assoc. of the Order of St. John of Jerusalem):

THE SURGEON'S POCKET-BOOK: an Essay on the Best Treatment of the Wounded in War; for which a Prize was awarded by Her Majesty the Empress of Germany. Specially adapted to the PUBLIC MEDICAL SERVICES. With 152 Illustrations, 16mo, roan. *Second Edition, Revised and Enlarged,* 7/6.

"Every Medical Officer is recommended to have the 'Surgeon's Pocket Book' by Surgeon-Major Porter, accessible to refresh his memory and fortify his judgment."—*Précis of Field-Service Medical Arrangements for Afghan War.*

"A complete *vade mecum* to guide the military surgeon in the field."—*British Medical Journal.*

"A capital little book . . . of the greatest practical value. . . . A surgeon with this Manual in his pocket becomes a man of resource at once."—*Westminster Review.*

"So fully illustrated that for LAY-READERS and AMBULANCE WORK it will prove eminently useful."—*Medical Times and Gazette.*

SCIENTIFIC MANUALS
BY
W. J. MACQUORN RANKINE, C.E., LL.D., F.R.S.,
Late Regius Professor of Civil Engineering in the University of Glasgow.

In Crown 8vo. Cloth.

I. RANKINE (Prof.): APPLIED MECHANICS: comprising the Principles of Statics and Cinematics, and Theory of Structures, Mechanism, and Machines. With numerous Diagrams. *Tenth Edition*, 12/6.

"Cannot fail to be adopted as a text-book. . . . The whole of the information is so admirably arranged that there is every facility for reference."—*Mining Journal.*

II. RANKINE (Prof.): CIVIL ENGINEERING: comprising Engineering Surveys, Earthwork, Foundations, Masonry, Carpentry, Metal-work, Roads, Railways, Canals, Rivers, Water-works, Harbours, &c. With numerous Tables and Illustrations. *Fourteenth Edition*, 16/.

"Far surpasses in merit every existing work of the kind. As a manual for the hands of the professional Civil Engineer it is sufficient and unrivalled, and even when we say this, we fall short of that high appreciation of Dr. Rankine's labours which we should like to express."—*The Engineer.*

III. RANKINE (Prof.): MACHINERY AND MILLWORK: comprising the Geometry, Motions, Work, Strength, Construction, and Objects of Machines, &c. Illustrated with nearly 300 Woodcuts. *Fifth Edition*, 12/6.

"Professor Rankine's 'Manual of Machinery and Millwork' fully maintains the high reputation which he enjoys as a scientific author; higher praise it is difficult to award to any book. It cannot fail to be a lantern to the feet of every engineer."—*The Engineer.*

IV. RANKINE (Prof.): THE STEAM ENGINE and OTHER PRIME MOVERS. With Diagram of the Mechanical Properties of Steam, Folding Plate, numerous Tables and Illustrations. *Tenth Edition*, 12/6.

V. RANKINE (Prof.): USEFUL RULES and TABLES for Engineers and others. With Appendix: TABLES, TESTS, and FORMULÆ for the use of ELECTRICAL ENGINEERS; comprising Submarine Electrical Engineering, Electric Lighting, and Transmission of Power. By ANDREW JAMIESON, C.E., F.R.S.E. *Sixth Edition*, 10/6.

"Undoubtedly the most useful collection of engineering data hitherto produced."—*Mining Journal.*
"Every Electrician will consult it with profit."—*Engineering.*

VI. RANKINE (Prof.): A MECHANICAL TEXT-BOOK. by Prof. MACQUORN RANKINE and E. F. BAMBER, C.E. With numerous Illustrations. *Second Edition*, 9/.

"The work, as a whole, is very complete, and likely to prove invaluable for furnishing a useful and reliable outline of the subjects treated of."—*Mining Journal.*

** THE MECHANICAL TEXT-BOOK forms a simple introduction to PROFESSOR RANKINE'S SERIES of MANUALS on ENGINEERING and MECHANICS.

PROF. RANKINE'S WORKS—(*Continued*).

VII. RANKINE (Prof.): MISCELLANEOUS
SCIENTIFIC PAPERS, from the Transactions and Proceedings of the Royal and other Scientific and Philosophical Societies, and the Scientific Journals. Royal 8vo. Cloth, 31/6.
PART I.—Papers relating to Temperature, Elasticity, and Expansion of Vapours, Liquids, and Solids.
PART II.—Papers on Energy and its Transformations.
PART III.—Papers on Wave-Forms, Propulsion of Vessels, &c.
With Memoir by P. G. TAIT, M.A., Prof. of Natural Philosophy in the University of Edinburgh. Edited by W. J. MILLAR, C.E., Secretary to the Institute of Engineers and Shipbuilders in Scotland. With fine Portrait on Steel, Plates, and Diagrams.

"No more enduring Memorial of Professor Rankine could be devised than the publication of these papers in an accessible form. . . . The Collection is most valuable on account of the nature of his discoveries, and the beauty and completeness of his analysis. . . . The Volume exceeds in importance any work in the same department published in our time."—*Architect.*

SCHWACKHÖFER (Professor Franz) and
BROWNE (Walter R., M.A., C.E., late Fellow of Trin. Coll., Camb.):
FUEL AND WATER: A Manual for Users of Steam and Water. Demy 8vo, with Numerous Illustrations, 9/. *Now Ready.*

GENERAL CONTENTS.

HEAT AND COMBUSTION.	FEED-WATER HEATERS.
FUEL, VARIETIES OF.	STEAM PIPES.
FIRING ARRANGEMENTS, FURNACE.	EXPLOSIONS.
FLUES, CHIMNEY.	WATER, COMPOSITION, PURIFICATION,
THE BOILER, CHOICE OF.	PREVENTION OF SCALE.
„ VARIETIES.	&c., &c.

SEATON (A. E., Lecturer on Marine Engineering
to the Royal Naval College, Greenwich, and Member of the Institute of Naval Architects):
A MANUAL OF MARINE ENGINEERING; Comprising the Designing, Construction, and Working of Marine Machinery. With numerous Illustrations. *Third Edition.* Demy 8vo. Cloth, 18/.

This Work has been prepared to supply the existing want of a Text-Book showing the application of Theoretical Principles to the DESIGN and CONSTRUCTION of MARINE MACHINERY as determined by the experience of leading Engineers, and carried out in the most recent successful practice. It is intended for use by Draughtsmen and practical Engineers, as well as by Instructors and Students in Marine Engineering, and is fully illustrated.

GENERAL CONTENTS.

I. PRINCIPLES OF MARINE PROPULSION.	III. DETAILS OF MARINE ENGINES.
II. PRINCIPLES OF STEAM ENGINEERING.	IV. PROPELLERS.
	V. BOILERS.
	VI. MISCELLANEOUS.

Opinions of the Press.

"The important subject of Marine Engineering is here treated with the thoroughness that it requires. No department has escaped attention. . . . Gives the results of much close study and practical work."—*Engineering.*

"By far the best MANUAL in existence. . . . Gives a complete account of the methods of solving, with the utmost possible economy, the problems before the Marine Engineer."—*Athenæum.*

"In the three-fold capacity of enabling a Student to learn how to design, construct, and work a modern Marine Steam-Engine, Mr. Seaton's Manual has no rival as regards comprehensiveness of purpose and lucidity of treatment."—*Times.*

"The Student, Draughtsman, and Engineer will find this work the most valuable Handbook of Reference on the Marine Engine now in existence."—*Marine Engineer.*

SHELTON (W. Vincent, Foreman to the Imperial
Ottoman Gun Factories, Constantinople):

THE MECHANIC'S GUIDE: A Hand-Book for Engineers and Artizans. With Copious Tables and Valuable Recipes for Practical Use. Illustrated. *Second Edition*. Crown 8vo. Cloth, 7/6.

GENERAL CONTENTS.

PART I.—Arithmetic.
PART II.—Geometry.
PART III.—Mensuration.
PART IV.—Velocities in Boring and Wheel-Gearing.
PART V.—Wheel and Screw Cutting.
PART VI.—Miscellaneous Subjects.
PART VII.—The Steam Engine.
PART VIII.—The Locomotive.

"The MECHANIC'S GUIDE will answer its purpose as completely as a whole series of elaborate text-books."—*Mining Journal*.

"Ought to have a place on the bookshelf of every Mechanic."—*Iron*.

"Much instruction is here given without pedantry or pretension."—*Builder*.

"A *sine quâ non* to every practical Mechanic."—*Railway Service Gazette*.

*** This Work is specially intended for Self-Teachers, and places before the Reader a concise and simple explanation of General Principles, together with Illustrations of their adaptation to Practical Purposes.

THOMSON (Spencer, M.D., L.R.C.S., Edinburgh,
and J. C. STEELE, M.D., of Guy's Hospital):

A DICTIONARY OF DOMESTIC MEDICINE AND HOUSEHOLD SURGERY. Thoroughly Revised and in part Re-Written by the Editors. With a Chapter on the Management of the Sick-room, and many Hints for the Diet and Comfort of Invalids. With many new Engravings. *Nineteenth Edition*. Royal 8vo. Cloth, 10/6.

WYLDE (James, formerly Lecturer on Natural
Philosophy at the Polytechnic):

THE MAGIC OF SCIENCE: A Manual of Easy and Amusing Scientific Experiments. With Steel Portrait of Faraday and many hundred Engravings. *Third Edition*. Crown 8vo. Cloth gilt and gilt edges, 5/.

"Of priceless value to furnish work for idle hands during the holidays. A thousand mysteries of Modern Science are here unfolded. We learn how to make Oxygen Gas, how to construct a Galvanic Battery, how to gild a Medal by Electro-Plating, or to reproduce one by Electrotyping, how to make a Microscope or take a Photograph; while the elements of Mechanics are explained so simply and clearly that the most unmechanical of minds must understand them. Such a work is deserving of the highest praise."—*The Graphic*.

"To those who need to be allured into the paths of Natural Science by witnessing the wonderful results that can be produced by well-contrived experiments, we do not know that we could recommend a more useful volume."—*Athenæum*.

EDUCATIONAL WORKS.

⁎ *Specimen Copies of all the Educational Works published by Messrs. Charles Griffin and Company may be seen at the Libraries of the College of Preceptors, South Kensington Museum, and Crystal Palace; also at the depôts of the Chief Educational Societies.*

BRYCE (Archibald Hamilton, D.C.L., LL.D.,
Senior Classical Moderator in the University of Dublin):
> THE WORKS OF VIRGIL. Text from HEYNE and WAGNER. English Notes, original, and selected from the leading German, American, and English Commentators. Illustrations from the antique. Complete in One Volume. *Fourteenth Edition.* Fcap 8vo. Cloth, 6/.
>
> Or, in Three Parts:
> Part I. BUCOLICS and GEORGICS, . . 2/6.
> Part II. THE ÆNEID, Books I.–VI., . 2/6.
> Part III. THE ÆNEID, Books VII.–XII., 2/6.

"Contains the pith of what has been written by the best scholars on the subject. . . . The notes comprise everything that the student can want."—*Athenæum.*

"The most complete, as well as elegant and correct edition of Virgil ever published in this country."—*Educational Times.*

"The best commentary on Virgil which a student can obtain."—*Scotsman.*

COBBETT (William): ENGLISH GRAMMAR,
in a Series of Letters, intended for the use of Schools and Young Persons in general. With an additional chapter on Pronunciation, by the Author's Son, JAMES PAUL COBBETT. *The only correct and authorised Edition.* Fcap 8vo. Cloth, 1/6.

"A new and cheapened edition of that most excellent of all English Grammars, William Cobbett's. It contains new copyright matter, as well as includes the equally amusing and instructive 'Six Lessons intended to prevent Statesmen from writing in an awkward manner.'"—*Atlas.*

COBBETT (William): A FRENCH GRAMMAR.
Fifteenth Edition. Fcap 8vo. Cloth, 3/6.

"Cobbett's 'French Grammar' comes out with perennial freshness. There are few grammars equal to it for those who are learning, or desirous of learning, French without a teacher. The work is excellently arranged, and in the present edition we note certain careful and wise revisions of the text."—*School Board Chronicle.*

"Business men commencing the study of French will find this treatise one of the best aids. . . . It is largely used on the Continent."—*Midland Counties Herald.*

COBBIN'S MANGNALL: MANGNALL'S
HISTORICAL AND MISCELLANEOUS QUESTIONS, for the use of Young People. By RICHMAL MANGNALL. Greatly enlarged and corrected, and continued to the present time, by INGRAM COBBIN, M.A. *Fifty-fourth Thousand. New Illustrated Edition.* 12mo. Cloth, 4/.

COLERIDGE (Samuel Taylor): A DISSER-
TATION ON THE SCIENCE OF METHOD. (*Encyclopædia
Metropolitana.*) With a Synopsis. *Ninth Edition.* Cr. 8vo. Cloth, 2/.

CRAIK'S ENGLISH LITERATURE.
A COMPENDIOUS HISTORY OF
ENGLISH LITERATURE AND OF THE ENGLISH LANGUAGE
FROM THE NORMAN CONQUEST. With numerous Specimens.
By GEORGE LILLIE CRAIK, LL.D., late Professor of History and
English Literature, Queen's College, Belfast. *New Edition.* In two
vols. Royal 8vo. Handsomely bound in cloth, 25/.

GENERAL CONTENTS.
INTRODUCTORY.

I.—THE NORMAN PERIOD—The Conquest.
II.—SECOND ENGLISH—Commonly called Semi-Saxon.
III.—THIRD ENGLISH—Mixed, or Compound English.
IV.—MIDDLE AND LATTER PART OF THE SEVENTEENTH CENTURY.
V.—THE CENTURY BETWEEN THE ENGLISH REVOLUTION AND
THE FRENCH REVOLUTION.
VI.—THE LATTER PART OF THE EIGHTEENTH CENTURY.
VII.—THE NINETEENTH CENTURY (*a*) THE LAST AGE OF THE
GEORGES.
(*b*) THE VICTORIAN AGE.

With numerous Excerpts and Specimens of Style.

"Anyone who will take the trouble to ascertain the fact, will find how completely even our great poets and other writers of the last generation have already faded from the view of the present, with the most numerous class of the educated and reading public. Scarcely anything is generally read except the publications of the day. YET NOTHING IS MORE CERTAIN THAN THAT NO TRUE CULTIVATION CAN BE SO ACQUIRED. This is the extreme case of that entire ignorance of history which has been affirmed, not with more point than truth, to leave a person always a child. . . . The present work combines the HISTORY OF THE LITERATURE with the HISTORY OF THE LANGUAGE. The scheme of the course and revolutions of the language which is followed here is extremely simple, and resting not upon arbitrary, but upon natural or real distinctions, gives us the only view of the subject that can claim to be regarded as of a scientific character."—*Extract from the Author's Preface.*

"Professor Craik's book going, as it does, through the whole history of the language, probably takes a place quite by itself. The great value of the book is its thorough comprehensiveness. It is always clear and straightforward, and deals not in theories but in facts."—*Saturday Review.*

"Professor Craik has succeeded in making a book more than usually agreeable."—*The Times.*

CRAIK (Prof.): A MANUAL OF ENGLISH
LITERATURE, for the use of Colleges, Schools, and Civil Service
Examinations. Selected from the larger work, by Dr. CRAIK. *Ninth
Edition.* With an Additional Section on Recent Literature, by HENRY
CRAIK, M.A., Author of "A Life of Swift." Crown 8vo. Cloth, 7/6.

"A Manual of English Literature from so experienced and well-read a scholar as Professor Craik needs no other recommendation than the mention of its existence."—*Spectator.*

"This augmented effort will, we doubt not, be received with decided approbation by those who are entitled to judge, and studied with much profit by those who want to learn. . . . If our young readers will give healthy perusal to Dr. Craik's work, they will greatly benefit by the wide and sound views he has placed before them."—*Athenæum.*

"The preparation of the NEW ISSUE has been entrusted to Mr. HENRY CRAIK, Senior Examiner in the Scotch Education Department, and well known in literary circles as the author of the latest and best Life of Swift. . . . A Series of TEST QUESTIONS is added, which must prove of great service to Students studying alone."—*Glasgow Herald.*

WORKS BY CHARLES T. CRUTTWELL, M.A.,
Fellow of Merton College, Oxford, and Head Master of Malvern College.

I.—A HISTORY OF ROMAN LITERA-
TURE: From the Earliest Period to the Times of the Antonines. *Third Edition.* Crown 8vo. Cloth, 8/6.

"Mr. CRUTTWELL has done a real service to all Students of the Latin Language and Literature. . . . Full of good scholarship and good criticism."—*Athenæum.*
"A most serviceable—indeed, indispensable—guide for the Student. . . . The 'general reader' will be both charmed and instructed."—*Saturday Review.*
"The Author undertakes to make Latin Literature interesting, and he has succeeded. There is not a dull page in the volume."—*Academy.*
"The great merit of the work is its fulness and accuracy."—*Guardian.*
"This elaborate and careful work, in every respect of high merit. Nothing at all equal to it has hitherto been published in England."—*British Quarterly Review.*

Companion Volume. Second Edition.

II.—SPECIMENS OF ROMAN LITERA-
TURE: From the Earliest Period to the Times of the Antonines. Passages from the Works of Latin Authors, Prose Writers, and Poets:
Part I.—ROMAN THOUGHT: Religion, Philosophy and Science, Art and Letters, 6/.
Part II.—ROMAN STYLE: Descriptive, Rhetorical, and Humorous Passages, 5/.
Or in One Volume complete, 10/6.
Edited by C. T. CRUTTWELL, M.A., Merton College, Oxford; and PEAKE BANTON, M.A., some time Scholar of Jesus College, Oxford.

"'Specimens of Roman Literature' marks a new era in the study of Latin."—*English Churchman.*
"Schoolmasters and tutors will be grateful for a volume which supplies them at once with passages of every shade of difficulty for testing the most different capacity, or which may be read with advantage in the higher forms of schools. There is no other book of the kind in this country which can be more safely recommended, either for its breadth, cheapness, or interest."—*Prof. Ellis in the "Academy."*
"A work which is not only useful but necessary. . . . The plan gives it a standing-ground of its own. . . . The sound judgment exercised in plan and selection calls for hearty commendation."—*Saturday Review.*
"It is hard to conceive a completer or handier repertory of specimens of Latin thought and style."—*Contemporary Review.*

** KEY to PART II., PERIOD II. (being a complete TRANSLATION of the 85 Passages composing the Section), by THOS. JOHNSTON, M.A., may now be had (by Tutors and Schoolmasters only) on application to the Publishers. Price 2/6.

CURRIE (Joseph, formerly Head Classical
Master of Glasgow Academy):
THE WORKS OF HORACE: Text from ORELLIUS. English Notes, original and selected, from the best Commentators. Illustrations from the antique. Complete in One Volume. Fcap 8vo. Cloth, 5/.
Or in Two Parts:
Part I.—CARMINA, ... 3/.
Part II.—SATIRES AND EPISTLES, . . 3/.
"The notes are excellent and exhaustive."—*Quarterly Journal of Education.*

EXTRACTS FROM CÆSAR'S COM-
MENTARIES; containing his description of Gaul, Britain, and Germany. With Notes, Vocabulary, &c. Adapted for Young Scholars. *Fourth Edition.* 18mo. Cloth, 1/6.

D'ORSEY (Rev. Alex. J. D., B.D., Corpus
Christi Coll., Cambridge, Lecturer at King's College, London):
SPELLING BY DICTATION: Progressive Exercises in English Orthography, for Schools and Civil Service Examinations. *Sixteenth Thousand.* 18mo. Cloth, 1/.

FLEMING (William, D.D., late Professor of
Moral Philosophy in the University of Glasgow):

THE VOCABULARY OF PHILOSOPHY: MENTAL, MORAL, AND METAPHYSICAL. With Quotations and References for the Use of Students. Revised and Edited by HENRY CALDERWOOD, LL.D., Professor of Moral Philosophy in the University of Edinburgh. *Third Edition, enlarged.* Crown 8vo. Cloth bevelled, 10/6.

"An admirable book. . . . In its present shape will be welcome, not only to Students, but to many who have long since passed out of the class of Students, popularly so called."—*Scotsman.*
"The additions by the Editor bear in their clear, concise, vigorous expression, the stamp of his powerful intellect. and thorough command of our language. More than ever, the work is now likely to have a prolonged and useful existence, and to facilitate the researches of those entering upon philosophic studies."—*Weekly Review.*
"A valuable addition to a Student's Library."—*Tablet.*

McBURNEY (Isaiah, LL.D.,): EXTRACTS
FROM OVID'S METAMORPHOSES. With Notes, Vocabulary, &c. Adapted for Young Scholars. *Third Edition.* 18mo. Cloth, 1/6.

MENTAL SCIENCE: S. T. COLERIDGE'S
celebrated Essay on METHOD; Archbishop WHATELY'S Treatises on LOGIC and RHETORIC. *Tenth Edition.* Crown 8vo. Cloth, 5/.

MILLER (W. Galbraith, M.A., LL.B., Lecturer
on Public Law, including Jurisprudence and International Law, in the University of Glasgow):
THE PHILOSOPHY OF LAW, LECTURES ON. Designed mainly as an Introduction to the Study of International Law. In 8vo. Handsome Cloth, 12/. *Now Ready.*

WORKS BY WILLIAM RAMSAY, M.A.,
Trinity College, Cambridge, late Professor of Humanity in the University of Glasgow.

A MANUAL OF ROMAN ANTIQUITIES.
For the use of Advanced Students. With Map, 130 Engravings, and very copious Index. *Twelfth Edition.* Crown 8vo. Cloth, 8/6.

"Comprises all the results of modern improved scholarship within a moderate compass."—*Athenæum.*

RAMSAY (Professor): AN ELEMENTARY
MANUAL OF ROMAN ANTIQUITIES. Adapted for Junior Classes. With numerous Illustrations. *Seventh Edition.* Crown 8vo. Cloth, 4/.

—— A MANUAL OF LATIN PROSODY,
Illustrated by Copious Examples and Critical Remarks. For the use of Advanced Students. *Sixth Edition.* Crown 8vo. Cloth, 5/.

"There is no other work on the subject worthy to compete with it."—*Athenæum.*

—— AN ELEMENTARY MANUAL OF
LATIN PROSODY. Adapted for Junior Classes. Crown 8vo. Cloth, 2/.

THE SCHOOL BOARD READERS:

A NEW SERIES OF STANDARD READING-BOOKS.
EDITED BY A FORMER H.M. INSPECTOR OF SCHOOLS.

Adopted by many School Boards throughout the Country.

Elementary Reader, Part I., 1d.	Standard III.,	.	.	9d.
,, ,, ,, II., 2d.	,, IV.,	.	.	1s. 0d.
Standard I., . . . 4d.	,, V.,	.	.	1s. 6d.
,, II., . . . 6d.	,, VI.,	.	.	2s. 0d.

Key to the Questions in Arithmetic in 2 Parts, each 6d.

*** Each Book of this Series contains within itself all that is necessary to fulfil the requirements of the Revised Code—viz., Reading, Spelling, and Dictation Lessons, together with Exercises in Arithmetic for the whole year. The paper, type, and binding are all that can be desired.

"THE BOOKS GENERALLY ARE VERY MUCH WHAT WE SHOULD DESIRE."—*Times.*
"The Series is DECIDEDLY ONE OF THE BEST that have yet appeared."—*Athenæum.*

THE SCHOOL BOARD MANUALS

ON THE SPECIFIC SUBJECTS OF THE REVISED CODE,
BY A FORMER H.M. INSPECTOR OF SCHOOLS,
Editor of the "School Board Readers."

64 pages, stiff wrapper. 6d.; neat cloth, 7d. each.

I.—ALGEBRA.
II.—ENGLISH HISTORY.
III.—GEOGRAPHY.
IV.—PHYSICAL GEOGRAPHY.
V.—ANIMAL PHYSIOLOGY. (Well Illustrated with good Engravings.)
VI.—BIBLE HISTORY. (Entirely free from any Denominational bias.)

"These simple and well-graduated Manuals, adapted to the requirements of the New Code, are the most elementary of elementary works, and extremely cheap. They are more useful as practical guide-books than most of the more expensive works."—*Standard.*

SENIOR (Nassau William, M.A., late Professor of Political Economy in the University of Oxford).
A TREATISE ON POLITICAL ECONOMY: the Science which treats of the Nature, the Production, and the Distribution of Wealth. *Sixth Edition.* Crown 8vo. Cloth. (*Encyclopædia Metropolitana*), 4/.

THOMSON (James): THE SEASONS. With an Introduction and Notes by ROBERT BELL, Editor of the "Annotated Series of British Poets." *Third Edition.* Fcap 8vo. Cloth, 1/6.
"An admirable introduction to the study of our English classics."

WHATELY (Archbishop): LOGIC—A Treatise on. With Synopsis and Index. (*Encyclopædia Metropolitana*), 3/.

—— RHETORIC—A Treatise on. With Synopsis and Index. (*Encyclopædia Metropolitana*), 3/6.

WYLDE (James): A MANUAL OF MATHEMATICS, Pure and Applied, 10/6.

WORKS IN GENERAL LITERATURE.

BELL (Robert, Editor of the " Annotated Series of British Poets"):

GOLDEN LEAVES FROM THE WORKS OF THE POETS AND PAINTERS. Illustrated by Sixty-four superb Engravings on Steel, after Paintings by DAVID ROBERTS, STANFIELD, LESLIE, STOTHARD, HAYDON, CATTERMOLE, NASMYTH, Sir THOMAS LAWRENCE, and many others, and engraved in the first style of Art by FINDEN, GREATBACH, LIGHTFOOT, &c. *Second Edition*, 4to. Cloth gilt, 21/.

"'Golden Leaves' is by far the most important book of the season. The Illustrations are really works of art, and the volume does credit to the arts of England."—*Saturday Review*.

"The Poems are selected with taste and judgment."—*Times*.

"The engravings are from drawings by Stothard, Newton, Danby, Leslie, and Turner, and it is needless to say how charming are many of the above here given."—*Athenæum*.

CHRISTISON (John): A COMPLETE SYSTEM OF INTEREST TABLES at 3, 4, 4½, and 5 per Cent.; Tables of Exchange or Commission, Profit and Loss, Discount, Clothiers', Malt, Spirit, and various other useful Tables. To which is prefixed the Mercantile Ready-Reckoner, containing Reckoning Tables from one thirty-second part of a penny to one pound. *New Edition. Greatly enlarged.* 12mo. Bound in leather, 4/6.

THE WORKS OF WILLIAM COBBETT.

THE ONLY AUTHORISED EDITIONS.

COBBETT (William): ADVICE TO YOUNG Men and (incidentally) to Young Women, in the Middle and Higher Ranks of Life. In a Series of Letters addressed to a Youth, a Bachelor, a Lover, a Husband, a Father, a Citizen, and a Subject. *New Edition. With admirable Portrait on Steel.* Fcap 8vo. Cloth, 2/6.

"Cobbett's great qualities were immense vigour, resource, energy, and courage, joined to a force of understanding, a degree of logical power, and above all a force of expression, which have rarely been equalled. . . . He was the most English of Englishmen."—*Saturday Review*.

"With all his faults, Cobbett's style is a continual refreshment to the lover of 'English undefiled.'"—*Pall Mall Gazette*.

WILLIAM COBBETT'S WORKS—*(Continued).*

COBBETT (Wm.): COTTAGE ECONOMY:
Containing information relative to the Brewing of Beer, Making of Bread, Keeping of Cows, Pigs, Bees, Poultry, &c.; and relative to other matters deemed useful in conducting the affairs of a Poor Man's Family. *Eighteenth Edition*, revised by the Author's Son. Fcap 8vo. Cloth, 2/6.

—— EDUCATIONAL WORKS. (See page 18.)

—— A LEGACY TO LABOURERS: An
Argument showing the Right of the Poor to Relief from the Land. With a Preface by the Author's Son, JOHN M. COBBETT, late M.P. for Oldham. *New Edition*. Fcap 8vo. Cloth, 1/6.

"The book cannot be too much studied just now."—*Nonconformist.*

"Cobbett was, perhaps, the ablest Political writer England ever produced, and his influence as a Liberal thinker is felt to this day. . . . It is a real treat to read his strong racy language."—*Public Opinion.*

—— A LEGACY TO PARSONS: Or, have the
Clergy of the Established Church an Equitable Right to Tithes and Church Property? *New Edition.* Fcap 8vo. Cloth, 1/6.

"The most powerful work of the greatest master of political controversy this country has ever produced."—*Pall Mall Gazette.*

COBBETT (Miss Anne): THE ENGLISH
HOUSEKEEPER; or, Manual of Domestic Management. Containing Advice on the conduct of Household Affairs and Practical Instructions, intended for the Use of Young Ladies who undertake the superintendence of their own Housekeeping. Fcap 8vo. Cloth, 3/6.

COOK'S VOYAGES. VOYAGES ROUND
THE WORLD, by Captain COOK. Illustrated with Maps and numerous Engravings. Two vols. Super-Royal 8vo. Cloth, 30/.

DALGAIRNS (Mrs.): THE PRACTICE OF
COOKERY, adapted to the business of Every-day Life. By Mrs. DALGAIRNS. *The best book for Scotch dishes.* About Fifty new Recipes have been added to the present Edition, but only such as the Author has had adequate means of ascertaining to be valuable. *Seventeenth Edition.* Fcap 8vo. Cloth. (*In preparation.*)

"This is by far the most complete and truly practical work which has yet appeared on the subject. It will be found an infallible 'Cook's Companion,' and a treasure of great price to the mistress of a family."—*Edinburgh Literary Journal.*

"We consider we have reason strongly to recommend Mrs. Dalgairns' as an economical, useful, and practical system of cookery, adapted to the wants of all families, from the tradesman's to the country gentleman's."—*Spectator.*

D'AUBIGNÉ (Dr. Merle): HISTORY OF THE REFORMATION. With the Author's latest additions and a new Preface. Many Woodcuts, and Twelve Engravings on Steel, illustrative of the life of MARTIN LUTHER, after LABOUCHÈRE. In one large volume. Demy 4to. Elegantly bound in cloth, 21/.

"In this edition the principal actors and scenes in the great drama of the Sixteenth Century are brought vividly before the eye of the reader, by the skill of the artist and Engraver."

DONALDSON (Joseph, Sergeant in the 94th Scots Regiment):
RECOLLECTIONS OF THE EVENTFUL LIFE OF A SOLDIER IN THE PENINSULA. *New Edition.* Fcap 8vo. Gilt sides and edges, 4/.

EARTH DELINEATED WITH PEN AND PENCIL (The): An Illustrated Record of Voyages, Travels, and Adventures all round the World. Illustrated with more than Two Hundred Engravings in the first style of Art, by the most eminent Artists, including several from the master-pencil of GUSTAVE DORÉ. Demy 4to, 750 pages. Very handsomely bound, 21/.

MRS. ELLIS'S CELEBRATED WORKS
On the INFLUENCE and CHARACTER of WOMEN.

THE ENGLISHWOMAN'S LIBRARY:

A Series of Moral and Descriptive Works. By Mrs. ELLIS. Small 8vo. Cloth, each volume, 2/6.

1. —THE WOMEN OF ENGLAND: Their Social Duties and Domestic Habits. *Thirty-ninth Thousand.*
2. —THE DAUGHTERS OF ENGLAND: Their Position in Society, Character, and Responsibilities. *Twentieth Thousand.*
3. —THE WIVES OF ENGLAND: Their Relative Duties, Domestic Influence, and Social Obligations. *Eighteenth Thousand.*
4. —THE MOTHERS OF ENGLAND: Their Influence and Responsibilities. *Twentieth Thousand.*
5. —FAMILY SECRETS; Or, Hints to make Home Happy. Three vols. *Twenty-third Thousand.*
6. —SUMMER AND WINTER IN THE PYRENEES. *Tenth Thousand.*
7. —TEMPER AND TEMPERAMENT; Or, Varieties of Character. Two vols. *Tenth Thousand.*
8. —PREVENTION BETTER THAN CURE; Or, the Moral Wants of the World we live in. *Twelfth Thousand.*
9. —HEARTS AND HOMES; Or, Social Distinctions. Three vols. *Tenth Thousand.*

THE EMERALD SERIES OF STANDARD AUTHORS.

Illustrated by Engravings on Steel, after STOTHARD, LESLIE, DAVID ROBERTS, STANFIELD, Sir THOMAS LAWRENCE, CATTERMOLE, &c., Fcap 8vo. Cloth, gilt.

> Particular attention is requested to this very beautiful series. The delicacy of the engravings, the excellence of the typography, and the quaint antique head and tail pieces, render them the most beautiful volumes ever issued from the press of this country, and now, unquestionably, the cheapest of their class.

BURNS' (Robert) SONGS AND BALLADS.
With an Introduction on the Character and Genius of Burns. By THOMAS CARLYLE. Carefully printed in antique type, and illustrated with Portrait and beautiful Engravings on Steel. *Second Thousand.* Cloth, gilt edges, 3/.

BYRON (Lord): CHILDE HAROLD'S PILGRIMAGE.
With Memoir by Professor SPALDING. Illustrated with Portrait and Engravings on Steel, by GREATBACH, MILLER, LIGHTFOOT, &c., from Paintings by CATTERMOLE, Sir T. LAWRENCE, H. HOWARD, and STOTHARD. Beautifully printed on toned paper. *Third Thousand.* Cloth, gilt edges, 3/.

CAMPBELL (Thomas): THE PLEASURES OF HOPE.
With Introductory Memoir by the Rev. CHARLES ROGERS, LL.D., and several Poems never before published. Illustrated with Portrait and Steel Engravings. *Second Thousand.* Cloth, gilt edges, 3/.

CHATTERTON'S (Thomas) POETICAL WORKS.
With an Original Memoir by FREDERICK MARTIN, and Portrait. Beautifully illustrated on Steel, and elegantly printed. *Fourth Thousand.* Cloth, gilt edges, 3/.

GOLDSMITH'S (Oliver) POETICAL WORKS.
With Memoir by Professor SPALDING. Exquisitely illustrated with Steel Engravings. *New Edition.* Printed on superior toned paper. *Seventh Thousand.* Cloth, gilt edges, 3/.

GRAY'S (Thomas) POETICAL WORKS.
With Life by the Rev. JOHN MITFORD, and Essay by the EARL of CARLISLE. With Portrait and numerous Engravings on Steel and Wood. Elegantly printed on toned paper. *Eton Edition, with the Latin Poems.* *Sixth Thousand.* Cloth, gilt edges, 5/.

HERBERT'S (George) POETICAL WORKS.
With Memoir by J. NICHOL, B.A., Oxon, Prof. of English Literature in the University of Glasgow. Edited by CHARLES COWDEN CLARKE. Antique headings to each page. *Second Thousand.* Cloth, gilt edges, 3/.

KEBLE (Rev. John): THE CHRISTIAN YEAR.
With Memoir by W. TEMPLE, Portrait, and Eight beautiful Engravings on Steel. *Second Thousand.*

Cloth, gilt edges,	5/.
Morocco, elegant,	10/6.
Malachite,	12/6.

The Emerald Series—(*Continued*).

POE'S (Edgar Allan) COMPLETE POETICAL
WORKS. Edited, with Memoir, by JAMES HANNAY. Full-page Illustrations after WEHNERT, WIER, &c. Toned paper. *Thirteenth Thousand.*
Cloth, gilt edges, ... 3/.
Malachite, ... 10/6.

Other volumes in preparation.

FINDEN'S FINE ART WORKS.

BEAUTIES OF MOORE: being a Series of
Portraits of his Principal Female Characters, from Paintings by eminent Artists, engraved in the highest style of Art by EDWARD FINDEN, with a Memoir of the Poet, and Descriptive Letterpress. Folio. Cloth gilt, 42/.

DRAWING-ROOM TABLE BOOK (The): a
Series of 31 highly-finished Steel Engravings, with descriptive Tales by Mrs. S. C. HALL, MARY HOWITT, and others. Folio. Cloth gilt, 21/.

GALLERY OF MODERN ART (The): a Series
of 31 highly-finished Steel Engravings, with descriptive Tales by Mrs. S. C. HALL, MARY HOWITT, and others. Folio. Cloth gilt, 21/.

FISHER'S READY-RECKONER. The best
in the World. *New Edition.* 18mo. Cloth, 1/6.

GILMER'S INTEREST TABLES; Tables for
Calculation of Interest, on any sum, for any number of days, at ½, 1, 1½, 2, 2½, 3, 3½, 4, 4½, 5 and 6 per Cent. By ROBERT GILMER. Corrected and enlarged. *Eleventh Edition.* 12mo. Cloth, 5/.

GRÆME (Elliott): BEETHOVEN: a Memoir.
With Portrait, Essay, and Remarks on the Pianoforte Sonatas, with Hints to Students, by DR. FERDINAND HILLER, of Cologne. *Second Edition slightly enlarged.* Crown 8vo. Cloth gilt, elegant, 5/.

"This elegant and interesting Memoir. . . . The newest, prettiest, and most readable sketch of the immortal Master of Music.—*Musical Standard.*
"A gracious and pleasant Memorial of the Centenary."—*Spectator.*
"This delightful little book — concise, sympathetic, judicious." — *Manchester Examiner.*
"We can, without reservation, recommend it as the most trustworthy and the pleasantest Memoir of Beethoven published in England."—*Observer.*
"A most readable volume, which ought to find a place in the library of every admirer of the great Tone-Poet."—*Edinburgh Daily Review.*

—— A NOVEL WITH TWO HEROES.
Second Edition. In 2 vols. Post 8vo. Cloth, 21/.

"A decided literary success."—*Athenæum.*
"Clever and amusing . . . above the average even of good novels . . . free from sensationalism, but full of interest . . . touches the deeper chords of life . . . delineation of character remarkably good."—*Spectator.*
"Superior in all respects to the common run of novels."—*Daily News.*
"A story of deep interest. . . . The dramatic scenes are powerful almost to painfulness in their intensity."—*Scotsman.*

HOGARTH: The Works of William Hogarth, in a Series of 150 Steel Engravings by the First Artists, with descriptive Letterpress by the Rev. JOHN TRUSLER, and Introductory Essay by JAMES HANNAY. Folio. Cloth, gilt edges, 52/6.

"The Philosopher who ever preached the sturdy English virtues which have made us what we are."

KNIGHT (Charles): PICTORIAL GALLERY OF THE USEFUL AND FINE ARTS. With Steel Engravings, and nearly 4,000 Woodcuts. Two vols. Folio. Cloth gilt, 42/.

────── **PICTORIAL MUSEUM OF ANIMATED NATURE.** Illustrated with 4,000 Woodcuts. Two vols. Folio. Cloth gilt, 35/.

MACKEY'S FREEMASONRY:
A LEXICON OF FREEMASONRY. Containing a definition of its Communicable Terms, Notices of its History, Traditions, and Antiquities, and an Account of all the Rites and Mysteries of the Ancient World. By ALBERT G. MACKEY, M.D., Secretary-General of the Supreme Council of the U.S., &c. *Seventh Edition*, thoroughly revised with APPENDIX by Michael C. Peck, Prov. Grand Secretary for N. and E. Yorkshire. Handsomely bound in cloth, 6/.

"Of MACKEY'S LEXICON it would be impossible to speak in too high terms; suffice it to say, that, in our opinion, it ought to be in the hands of every Mason who would thoroughly understand and master our noble Science. . . . No Masonic Lodge or Library should be without a copy of this most useful work."—*Masonic News*.

HENRY MAYHEW'S CELEBRATED WORK ON THE STREET-FOLK OF LONDON.

LONDON LABOUR AND THE LONDON POOR: A Cyclopædia of the Condition and Earnings of *those that will work and those that cannot work*. By HENRY MAYHEW. With many full-page Illustrations from Photographs. In three vols. Demy 8vo. Cloth. Each vol. 4/6.

"Every page of the work is full of valuable information, laid down in so interesting a manner that the reader can never tire."—*Illustrated London News*.

"Mr. Henry Mayhew's famous record of the habits, earnings, and sufferings of the London poor."—*Lloyd's Weekly London Newspaper*.

"This remarkable book, in which Mr. Mayhew gave the better classes their first real insight into the habits, modes of livelihood, and current of thought of the London poor."—*The Patriot*.

The Extra Volume.

LONDON LABOUR AND THE LONDON POOR: *Those that will not work*. Comprising the Non-workers, by HENRY MAYHEW; Prostitutes, by BRACEBRIDGE HEMYNG; Thieves, by JOHN BINNY; Beggars, by ANDREW HALLIDAY. With an Introductory Essay on the Agencies at Present in Operation in the Metropolis for the Suppression of Crime and Vice, by the Rev. WILLIAM TUCKNISS, B.A., Chaplain to the Society for the Rescue of Young Women and Children. With Illustrations of Scenes and Localities. In one large vol. Royal 8vo. Cloth, 10/6.

"The work is full of interesting matter for the casual reader, while the philanthropist and the philosopher will find details of the greatest import."—*City Press*.

Mr. Mayhew's London Labour—(*Continued*).

Companion volume to the preceding.
THE CRIMINAL PRISONS OF LONDON,
and Scenes of Prison Life. By HENRY MAYHEW and JOHN BINNY. Illustrated by nearly two hundred Engravings on Wood, principally from Photographs. In one large vol. Imperial 8vo. Cloth, 10/6.

"This volume concludes Mr. Henry Mayhew's account of his researches into the crime and poverty of London. The amount of labour of one kind or other, which the whole series of his publications represents, is something almost incalculable."—*Literary Budget.*

*** This celebrated Record of Investigations into the condition of the Poor of the Metropolis, undertaken from philanthropic motives by Mr. HENRY MAYHEW, first gave the wealthier classes of England some idea of the state of Heathenism, Degradation, and Misery in which multitudes of their poorer brethren languished. His revelations created, at the time of their appearance, universal horror and excitement—that a nation, professedly *Christian*, should have in its midst a vast population, so sunk in ignorance, vice, and very hatred of Religion, was deemed incredible, until further examination established the truth of the statements advanced. The result is well known. The London of Mr. MAYHEW will, happily, soon exist only in his pages. To those who would appreciate the efforts already made among the ranks which recruit our "dangerous" classes, and who would learn what yet remains to be done, the work will afford enlightenment, not unmingled with surprise.

MILLER (Thomas, Author of "Pleasures of a Country Life," &c.):
THE LANGUAGE OF FLOWERS. With Eight beautifully-coloured Floral Plates. Fcap 8vo. Cloth, gilt edges. *Fourteenth Thousand*, ... 3/6.
    Morocco, ... 7/6.

"A book
In which thou wilt find many a lovely saying
About the leaves and flowers."—KEATS.

——— THE LANGUAGE OF FLOWERS
Abridged from the larger work by THOMAS MILLER. With coloured Frontispiece. *Cheap Edition.* Limp cloth, 6d.

POE'S (Edgar Allan) COMPLETE POETICAL
WORKS. Edited, with Memoir, by JAMES HANNAY. Full-page Illustrations after WEHNERT, WEIR, and others. In paper wrapper. Illustrated, 1/6.

POETRY OF THE YEAR: Or, Pastorals from
our Poets, illustrative of the Seasons. With Chromo-Lithographs from Drawings after BIRKET FOSTER, R.A., S. CRESWICK, R.A., DAVID COX, HARRISON WEIR, E.V.B., and others. *New Edition.* Toned paper. Cloth gilt, elegant, 16/.

RAPHAEL: THE CARTOONS OF
RAPHAEL. Engraved on Steel in the first style of Art by G. GREATBACH, after the Originals at South Kensington. With Memoir, Portrait of RAPHAEL, as painted by himself, and Fac-simile of his Autograph. Folio. Elegantly bound in cloth, 10/6.

"Forms a handsome volume."—*Art and Letters.*

SCHILLER'S MAID OF ORLEANS: (*Die Jungfrau von Orleans*). Rendered into English by LEWIS FILMORE, translator of GOETHE'S FAUST. With admirable Portrait of SCHILLER, engraved on Steel by ADLARD, and Introductory Notes. In Crown 8vo. Toned paper. Cloth, elegant, gilt edges, 2/6.

> "Mr. Filmore's excellent translation deserves to be read by all."—*Northern Whig.*
> "The drama has found in Mr Filmore a faithful and sympathetic translator."—*Public Opinion.*

SHAKSPEARE: THE FAMILY. The Dramatic Works of WILLIAM SHAKSPEARE, edited and expressly adapted for Home and School Use. By THOMAS BOWDLER, F.R.S. With Twelve beautiful Illustrations on Steel. *New Edition.* Crown 8vo.

Cloth, gilt, ... 10/6.
Morocco antique, ... 17/6.

*** *This unique Edition of the great dramatist is admirably suited for home use; while objectionable phrases have been expurgated, no rash liberties have been taken with the text.*

> "It is quite undeniable that there are many passages in Shakspeare which a father could not read aloud to his children—a brother to his sister—or a gentleman to a lady; and every one almost must have felt or witnessed the extreme awkwardness, and even distress, that arises from suddenly stumbling upon such expressions. . . . Those who recollect such scenes must all rejoice that Mr. BOWDLER has provided a security against their recurrence. . . . This purification has been accomplished with surprisingly little loss, either of weight or value; the base alloy in the pure metal of Shakspeare has been found to amount to an inconceivably small proportion. . . . It has in general been found easy to extirpate the offensive expressions of our great poet without any injury to the context, or any visible scar or blank in the composition. They turn out to be not so much cankers in the flowers, as weeds that have sprung up by their side—not flaws in the metal, but impurities that have gathered on its surface—and, so far from being missed, on their removal the work generally appears more natural and harmonious without them."—*Lord Jeffrey in the Edinburgh Review.*

SHAKSPEARE'S DRAMATIC & POETICAL WORKS. Revised from the Original Editions, with a Memoir and Essay on his Genius by BARRY CORNWALL; and Annotations and Introductory Remarks on his Plays, by R. H. HORNE, and other eminent writers. With numerous Woodcut Illustrations and full-page Steel Engravings by KENNY MEADOWS. *Tenth Edition.* Three vols. Super-royal 8vo. Cloth, gilt, 42/.

SHAKSPEARE'S WORKS. Edited by T. O. HALLIWELL, F.R.S., F.S.A. With Historical Introductions, Notes, Explanatory and Critical, and a Series of Portraits on Steel. Three vols. Royal 8vo. Cloth gilt, 50/.

SOUTHGATE (Mrs. Henry): THE CHRISTIAN LIFE: Thoughts in Prose and Verse from the Best Writers of all Ages. Selected and Arranged for Every Day in the Year. 8vo. Cloth Elegant, ... 5/.
Morocco Antique, . 10/6.

MR. SOUTHGATE'S WORKS.

"No one who is in the habit of writing and speaking much on a variety of subjects can afford to dispense with Mr. SOUTHGATE'S WORKS."—*Glasgow News.*

FIRST SERIES—THIRTY-SECOND EDITION. SECOND SERIES—
EIGHTH EDITION.

MANY THOUGHTS OF MANY MINDS:

Selections and Quotations from the best Authors. Compiled and
Analytically Arranged by

HENRY SOUTHGATE.

In Square 8vo, elegantly printed on Toned Paper.

Presentation Edition, Cloth and Gold, ...	Each Vol. 12/6.
Library Edition, Roxburghe, ...	,, 14/.
Ditto, Morocco Antique, ...	,, 21/.

Each Series complete in itself, and sold separately.

"The produce of years of research."—*Examiner.*
"A MAGNIFICENT GIFT-BOOK, appropriate to all times and seasons."—*Freemasons' Magazine.*
"Not so much a book as a library."—*Patriot.*
"Preachers and Public Speakers will find that the work has special uses for them."—*Edinburgh Daily Review.*

BY THE SAME AUTHOR.

Now Ready, Third Edition.

SUGGESTIVE THOUGHTS ON RELIGIOUS SUBJECTS:

A Dictionary of Quotations and Selected Passages from nearly 1,000 of
the best Writers, Ancient and Modern.
Compiled and Analytically Arranged by HENRY SOUTHGATE. In
Square 8vo, elegantly printed on toned paper.

Presentation Edition, Cloth Elegant, ...	10/6.
Library Edition, Roxburghe, ...	12/.
Ditto, Morocco Antique, ...	20/.

"The topics treated of are as wide as our Christianity itself: the writers quoted from, of every Section of the one Catholic Church of JESUS CHRIST."—*Author's Preface.*
"This is another of Mr. Southgate's most valuable volumes. . . . The mission which the Author is so successfully prosecuting in literature is not only highly beneficial, but necessary in this age. . . . If men are to make any acquaintance at all with the great minds of the world, they can only do so with the means which our Author supplies.—*Homilist.*
"A casket of gems."—*English Churchman.*
"Mr. Southgate's work has been compiled with a great deal of judgment, and it will, I trust, be extensively useful."—*Rev. Canon Liddon, D.D., D.C.L.*
"Many a busy Christian teacher will be thankful to Mr. Southgate for having unearthed so many rich gems of thought ; while many outside the ministerial circle will obtain stimulus, encouragement, consolation, and counsel, within the pages of this handsome volume."—*Nonconformist.*
"The special value of this most admirable compilation is discovered, when attention is concentrated on a particular subject, or series of subjects, as illustrated by the various and often brilliant lights shed by passages selected from the best authors in all ages. . . . A most valuable book of reference."- *Edinburgh Daily Review.*
"Mr. SOUTHGATE is an indefatigable labourer in a field which he has made peculiarly his own. . . . The labour expended on 'Suggestive Thoughts' must have been immense, and the result is as nearly perfect as human fallibility can make it. . . . Apart from the selections it contains, the book is of value as an index to theological writings. As a model of judicious, logical, and suggestive treatment of a subject, we may refer our readers to the manner in which the subject 'JESUS CHRIST' is arranged and illustrated in 'Suggestive Thoughts.' "—*Glasgow News.*

THE SHILLING MANUALS.

By JOHN TIMBS, F.S.A.,
Author of "The Curiosities of London," &c.

A Series of Hand-Books, containing Facts and Anecdotes interesting to all Readers. *Second Edition.* Fcap 8vo. Bound in neat cloth.

Price One Shilling each.

I.—TIMBS' CHARACTERISTICS OF EMINENT MEN.

II.—TIMBS' CURIOSITIES OF ANIMAL AND VEGETABLE LIFE.

III.—TIMBS' ODDITIES OF HISTORY AND STRANGE STORIES FOR ALL CLASSES.

IV.—TIMBS' ONE THOUSAND DOMESTIC HINTS on the Choice of Provisions, Cookery, and Housekeeping; new Inventions and Improvements; and various branches of Household Management.

V.—TIMBS' POPULAR SCIENCE: Recent Researches on the Sun, Moon, Stars, and Meteors; the Earth; Phenomena of Life, Sight, and Sound; Inventions and Discoveries.

VI.—TIMBS' THOUGHTS FOR TIMES AND SEASONS.

Opinions of the Press on the Series.

"It is difficult to determine which of these volumes is the most attractive. Will be found equally enjoyable on a railway journey or by the fireside."—*Mining Journal.*

"These additions to the Library, produced by Mr. Timbs' industry and ability, are useful, and in his pages many a hint and suggestion, and many a fact of importance is stored up that would otherwise have been lost to the public."—*Builder.*

"Capital little books of about a hundred pages each, wherein the indefatigable Author is seen at his best."—*Mechanic's Magazine.*

"Extremely interesting volumes."—*Evening Standard.*

"Amusing, instructive, and interesting. . . . As food for thought and pleasant reading, we can heartily recommend the 'Shilling Manuals.'"—*Birmingham Daily Gazette.*

TIMBS (John, F.S.A.): PLEASANT HALF-HOURS FOR THE FAMILY CIRCLE. Containing Popular Science, One Thousand Domestic Hints, Thoughts for Times and Seasons, Oddities of History, and Characteristics of Great Men. *Second Edition.* Fcap 8vo. Cloth gilt, and gilt edges, 5/.

"Contains a wealth of useful reading of the greatest possible variety."—*Plymouth Mercury.*

WANDERINGS IN EVERY CLIME; Or,
Voyages, Travels, and Adventures All Round the World. Edited by W. F. AINSWORTH, F.R.G.S., F.S.A., &c., and embellished with upwards of Two Hundred Illustrations by the first Artists, including several from the master-pencil of GUSTAVE DORÉ. Demy 4to. 800 pages. Cloth and gold, bevelled boards, 21/.

INDEX.

	PAGE
AINSWORTH (W. F.), Earth Delineated,	25
—— Wanderings in Every Clime,	32
AITKEN (W., M.D.), Science and Practice of Medicine,	6
—— Outlines,	7
—— Growth of the Recruit,	7
ANSTED (Prof.), Geology,	9
—— Inanimate Creation,	7
BAIRD (Prof.), Student's Natural History,	7
BELL (R.), Golden Leaves,	23
BLYTH (A. W.), Hygiène and Public Health,	8
—— Foods and Poisons,	8
BROUGHAM (Lord), Paley's Natural Theology,	5
BROWNE (W. R.), Student's Mechanics,	7
—— Foundations of Mechanics,	7
—— Fuel and Water,	16
BRYCE (A. H.), Works of Virgil,	18
—— Works of Virgil (in Parts),	18
BUNYAN'S Pilgrim's Progress (Mason),	1
—————— Do. (Maguire),	1
—— Select Works,	1
CHEEVER'S (Dr.), Religious and Moral Anecdotes,	1
CHRISTISON (J.), Interest Tables,	23
CIRCLE OF THE SCIENCES, 9 vols.,	9
—— Treatises,	9
COBBETT (Wm.), Advice to Young Men,	23
—— Cottage Economy,	24
—— English Grammar,	18
—— French do.,	18
—— Legacy to Labourers,	24
—— Do. Parsons,	24
COBBIN'S Mangnall's Question,	18
COLERIDGE on Method,	19
COOK (Captain), Voyages of,	24
CRAIK (G.), History of English Literature,	19
—— Manual of do.	19
CRUDEN'S CONCORDANCE, by Eadie,	3
—— by Youngman,	2
CRUTTWELL'S History of Roman Literature,	20
———— Specimens of do.,	20
—— Do. do. (in Parts),	20
CURRIE (J.), Works of Horace,	20
—— Do. (in Parts),	20
—— Cæsar's Commentaries,	20
DALGAIRN'S COOKERY,	24
DALLAS (Prof.), Animal Creation,	10
D'AUBIGNE'S History of the Reformation,	25
DICK (Dr.), Celestial Scenery,	2
—— Christian Philosopher,	2
DONALDSON (Jas.), Eventful Life of a Soldier,	25
D'ORSEY (A. J.), Spelling by Dictation,	20
DOUGLAS (J. C.), Manual of Telegraph Construction,	10
DUPRE AND HAKE, Practical Chemistry,	10
EADIE (Rev. Dr.), Biblical Cyclopædia,	3
—— Cruden's Concordance,	3
—— Classified Bible,	3
—— Ecclesiastical Cyclopædia,	3
—— Dictionary of Bible,	3
ELLIS (Mrs.), Englishwoman's Library,	25
EMERALD SERIES OF STANDARD AUTHORS,	26
FINDEN'S FINE-ART WORKS,	27
FISHER'S READY-RECKONER,	27
FLEMING (Prof.), Vocabulary of Philosophy,	21
FOSTER (Chas.), Story of the Bible,	4
GEDGE (Rev. J. W.), Bible History,	1
GILMER (R.), Interest Tables,	27

	PAGE
GORE (G.), Electro-deposition,	9
GRÆME (Elliott), Beethoven,	27
—— Novel with Two Heroes,	27
GRIFFIN (J. J.), Chemical Recreations,	10
—— Do. (in Parts),	10
GURDEN (R.), Traverse Tables,	11
HARRIS (Rev. Dr.), Altar of Household,	1
HENRY (M.), Commentary on the Bible,	2
HOGARTH, Works of,	28
JAMIESON (A.), Manual of the Steam Engine	11
KEBLE'S CHRISTIAN YEAR, 4to,	2
—— Do., Fcap,	2
KITTO (Rev. Dr.), The Holy Land,	4
—— Pictorial Sunday Book,	4
KNIGHT (Charles), Pictorial Gallery,	28
—— Do. Museum,	28
LEARED (Dr.), Imperfect Digestion,	11
LINN (Dr.), On the Teeth,	12
LONGMORE (Prof.), Sanitary Contrasts,	12
M'BURNEY (Dr.), Ovid's Metamorphoses,	21
MACKEY (A. G.), Lexicon of Freemasonry,	28
M'NAB (Dr.), Manual of Botany,	12
MAYHEW (H.), London Labour,	28
MENTAL SCIENCE (Coleridge and Whately),	21
MILLER (T.), Language of Flowers,	29
MILLER (W. G.), Philosophy of Law,	21
MOFFITT (Dr.), Instruction for Attendants on Wounded,	12
MUNRO AND JAMIESON'S Electrical Pocket-Book,	12
NAPIER (Jas.), Dyeing and Dyeing Receipts,	13
—— Electro-Metallurgy,	13
PHILLIPS (John), Manual of Geology,	13
PHILLIPS (J. A.), Elements of Metallurgy,	14
POE (Edgar), Poetical Works of,	27
POETRY OF THE YEAR,	29
PORTER (Surg.-Maj.), Surgeon's Pocket-Book,	14
RAGG (Rev. T.), Creation's Testimony,	5
RAMSAY (Prof.), Roman Antiquities,	21
—— Do. Elemy.,	21
—— Latin Prosody,	21
—— Do. Elemy.,	21
RANKINE'S ENGINEERING WORKS,	15, 16
RAPHAEL'S CARTOONS,	29
RELIGIONS OF THE WORLD,	5
SCHILLER'S MAID OF ORLEANS,	30
SCHOOL BOARD MANUALS,	22
—— READERS,	22
SCOTT (Rev. Thos.), Commentary on the Bible,	2
SEATON (A. E.), Marine Engineering,	16
SENIOR (Prof.), Political Economy,	22
SHAKESPERE, Bowdler's Family,	30
—— Barry Cornwall's,	30
—— Halliwell's,	30
SHELTON (W. V.), Mechanic's Guide,	17
SOUTHGATE (H.), Many Thoughts of Many Minds,	31
—— Suggestive Thoughts,	31
—— (Mrs.), Christian Life,	5
THOMSON (Dr. Spencer), Domestic Medicine,	17
THOMSON'S SEASONS,	22
TIMBS' (John), Shilling Manuals,	32
—— Pleasant Half Hours,	32
WHATELY (Archbishop), Logic,	22
—— Rhetoric,	22
WORDS AND WORKS OF OUR BLESSED LORD,	5
WYLDE (Jas.), Magic of Science,	17
—— Manual of Mathematics,	22

FIRST SERIES.—THIRTY-SECOND EDITION.
SECOND SERIES.—EIGHTH EDITION.

MANY THOUGHTS OF MANY MINDS:
A TREASURY OF REFERENCE,
Consisting of Selections from the Writings of the most Celebrated Authors.
FIRST AND SECOND SERIES. COMPILED AND ANALYTICALLY ARRANGED
By HENRY SOUTHGATE.

In Square 8vo, elegantly printed on Toned Paper.

Presentation Edition, Cloth and Gold, .	12s. 6d. each Volume.	
Library Edition, Half-Bound, Roxburghe, .	14s.	,,
Do., Morocco Antique, . .	21s.	,,

Each Series is Complete in itself, and sold separately.

"'MANY THOUGHTS,' &c., are evidently the produce of years of research. We look up any subject under the sun, and are pretty sure to find something that has been said—generally *well said*—upon it."—*Examiner.*

"Many beautiful examples of thought and style are to be found among the selections."—*Leader.*

"There can be little doubt that it is destined to take a high place among books of this class."—*Notes and Queries.*

"A treasure to every reader who may be fortunate enough to possess it. Its perusal is like inhaling essences; we have the cream only of the great authors quoted. Here all are seeds or gems."—*English Journal of Education.*

"Mr. Southgate's reading will be found to extend over nearly the whole known field of literature, ancient and modern."—*Gentleman's Magazine.*

"Here is matter suited to all tastes, and illustrative of all opinions—morals, politics, philosophy, and solid information. We have no hesitation in pronouncing it one of the most important books of the season. Credit is due to the publishers for the elegance with which the work is got up, and for the extreme beauty and correctness of the typography."—*Morning Chronicle.*

"Of the numerous volumes of the kind, we do not remember having met with one in which the selection was more judicious, or the accumulation of treasures so truly wonderful."—*Morning Herald.*

"Mr. Southgate appears to have ransacked every nook and corner for gems of thought."—*Allen's Indian Mail.*

"The selection of the extracts has been made with taste, judgment, and official nicety."—*Morning Post.*

"This is a wondrous book, and contains a great many gems of thought."—*Daily News.*

"As a work of reference, it will be an acquisition to any man's library."—*Publishers' Circular.*

"This volume contains more gems of thought, refined sentiments, noble axioms, and extractable sentences, than have ever before been brought together in our language."—*The Field.*

"Will be found to be worth its weight in gold by literary men."—*The Builder.*

"All that the poet has described of the beautiful in nature and art; all the wit that has flashed from pregnant minds; all the axioms of experience, the collected wisdom of philosopher and sage, are garnered into one heap of useful and well-arranged instruction and amusement."—*The Era.*

"The mind of almost all nations and ages of the world is recorded here."—*John Bull.*

"This is not a law-book; but, departing from our usual practice, we notice it because it is likely to be very useful to lawyers."—*Law Times.*

"The collection will prove a mine, rich and inexhaustible, to those in search of a quotation."—*Art Journal.*

"There is not, as we have reason to know, a single trashy sentence in this volume. Open where we may, every page is laden with the wealth of profoundest thought, and all aglow with the loftiest inspirations of genius. To take this book into our hands is like sitting down to a grand conversazione with the greatest thinkers of all ages."—*Star.*

"The work of Mr. Southgate far outstrips all others of its kind. To the clergyman, the author, the artist, and the essayist, 'Many Thoughts of Many Minds' cannot fail to render almost incalculable service."—*Edinburgh Mercury.*

"We have no hesitation whatever in describing Mr. Southgate's as the very best book of the class. There is positively nothing of the kind in the language that will bear a moment's comparison with it."—*Manchester Weekly Advertiser.*

"There is no mood in which we can take it up without deriving from it instruction, consolation, and amusement. We heartily thank Mr. Southgate for a book which we shall regard as one of our best friends and companions."—*Cambridge Chronicle.*

"This work possesses the merit of being a magnificent gift-book, appropriate to all times and seasons—a book calculated to be of use to the scholar, the divine, or the public man."—*Freemasons' Magazine.*

"It is not so much a book as a library of quotations."—*Patriot.*

"The quotations abound in that *thought* which is the mainspring of mental exercise."—*Liverpool Courier.*

"For purposes of apposite quotation it cannot be surpassed."—*Bristol Times.*

"It is impossible to pick out a single passage in the work which does not, upon the face of it, justify its selection by its intrinsic merit."—*Dorset Chronicle.*

"We are not surprised that a Second Series of this work should have been called for. Mr. Southgate has the catholic tastes desirable in a good editor. Preachers and public speakers will find that it has special uses for them."—*Edinburgh Daily Review.*

"The SECOND SERIES fully sustains the deserved reputation of the First."—*John Bull.*

LONDON: CHARLES GRIFFIN & COMPANY.

www.ingramcontent.com/pod-product-compliance
Lightning Source LLC
Chambersburg PA
CBHW031940230426
43672CB00010B/1985